PRE-U PURE MATHEMATICS

大学预科数学

主编

张 弘 (Zhang Hong) 江苏大学

【罗马尼亚】保罗·乔治斯库 (Paul Georgescu) 罗马尼亚雅西技术大学 Technical University of Iasi, Romania

副主编

张 来 扬州大学

朱 劼 常州大学

参编

Prince Harvim

Takuriramunashe Famba

Wilson Osafo Apeanti

Akuamoah Worlanyo Saviour

Kwasi Boateng

David Yaro

江苏大学出版社
JIANGSU UNIVERSITY PRESS

镇 江

图书在版编目(CIP)数据

大学预科数学 = Pre-U Pure Mathematics：英文 / 张弘，(罗)保罗 · 乔治斯库(Paul Georgescu) 主编 . —镇江：江苏大学出版社，2018.7
ISBN 978-7-5684-0852-3

Ⅰ.①大… Ⅱ.①张… ②保… Ⅲ.①数学－高等学校－教材－英文 Ⅳ.①O1

中国版本图书馆 CIP 数据核字(2018)第 142455 号

大学预科数学
Pre-U Pure Mathematics

主　　编/张　弘　【罗马尼亚】保罗 · 乔治斯库(Paul Georgescu)
责任编辑/吴昌兴
出版发行/江苏大学出版社
地　　址/江苏省镇江市梦溪园巷 30 号(邮编：212003)
电　　话/0511-84446464(传真)
网　　址/http：//press.ujs.edu.cn
排　　版/镇江文苑制版印刷有限责任公司
印　　刷/虎彩印艺股份有限公司
开　　本/787 mm×1 092 mm　1/16
印　　张/14.25
字　　数/462 千字
版　　次/2018 年 7 月第 1 版　2018 年 7 月第 1 次印刷
书　　号/ISBN 978-7-5684-0852-3
定　　价/45.00 元

如有印装质量问题请与本社营销部联系(电话：0511-84440882)

PREFACE

Pre-U Pure Mathematics is a textbook which is specifically aimed at the international students in the pre-university foundation school of Jiangsu University. It has been developed following an extensive period of research and consultation with a large number of teachers and students, several of its chapters being already tried and tested in the foundation school, and it is intended to be usable both as a classroom resource and as a textbook to support self-motivated learners away for classroom.

This textbook covers all the requirements for Pure Mathematics from the latest pre-university level specifications and course requirements for the international students studying in China. From the very beginning, it has been conceived to be as clear and helpful as possible. It uses a concrete approach, introducing new concepts or revisiting older ones via extensive sets of worked examples and clear figures and diagrams, aiming at diminishing the discomfort which is often caused at this level by a more abstract approach and at developing a solid conceptual basis on which deeper understanding can be built upon.

The presentation builds on a minimal amount of required knowledge, attempting to provide comprehensive explanations of all topics covered in a friendly, conversational style. The presentation is split into smaller units or modules, clearly labeled as such, each dealing with a single aspect, in clear progression. This makes for easier knowledge assimilation, one step at a time, and helps the reader proceed at a steady pace, while isolating the possible difficulties.

Each chapter contains a comprehensive summary, which lets students review the key concepts and methods, and numerous exercises to reinforce understanding via problem-solving. Remarks and margin notes provide supporting commentary on key definitions, topics and formulae, aiming at clarifying difficult aspects and sometimes at providing mnemonics for remembering formulae and key steps of approaches.

Mathematical knowledge can be delivered in various manners, but ultimately under-standing comes from within the learner, when the individual participates, explores, discovers and makes connections. We hope that this textbook will not only deliver content, but will also be a worthy companion and a helpful guide on this road to understanding.

The Authors

CONTENTS

Chapter 1 Algebra I

1.1 Sets and numbers

1.1.1 Set theory

Definition 1.1.1—Set. A **set** is simply a collection of objects.

There are two ways of describing the members of a set. One way is by **intensional definition**, using a rule or a semantic description.

■ **Example 1.1** (1) A is the set whose members are positive integers that is less than 10.

(2) B is the set of colors of the flag of China.

■

The second way is by **extensional definition**, enclosing the list of members in curly brackets.

■ **Example 1.2** (1) $\{1,2,3\}$ is the set consisting of the objects 1,2,3.

(2) {Nike, Adidas, Reebok, Li Ning, Under Armour} is the set consisting of the objects Nike, Adidas, Reebok, Li Ning, Under Armour.

(3) $\{\{1\},2\}$ is the set consisting of two objects, one being the set $\{1\}$ and the other being the number 2.

■

Ⓡ The order in which the objects are written doesn't matter. For instance, the set

{Nike, Adidas, Reebok, Li Ning, Under Armour}

is the same set as

{Reebok, Li Ning, Nike, Under Armour, Adidas}.

Often, however, this is not a convenient way to describe our set. For example, the set consisting of all the people who live in China is, for most purposes, best described by precisely this phrase. Describing this set in name list form is unlikely to be useful. As another example, the set of all real numbers whose square is greater than 10 is neatly described by the notation

$$\{x \mid x \text{ is a real number and } x^2 > 10\}.$$

This is to be read "the set of all x such that x is a real number and $x^2 > 10$". The symbol (vertical bar) "|" is the "such that" part of the phrase.

Definition 1.1.2—Element. The objects that are in the set are known as **elements or members** of the set.

Notation 1.1 If x is an element of a set A, we then write $x \in A$. This can also be read as "x is a member of a set A" or "x belongs to A" or "x is in A". The notation $x \notin A$ means that x is not a member of A.

Ⓡ The objects that make up a set can be anything: numbers, people, letters of the alphabet, other sets, and so on.

■ **Example 1.3**

$1 \in \{1,2,3\}$, but $4 \notin \{1,2,3\}$.

■

■ **Example 1.4**

$\{1,3,5\}$ is the set consisting the objects 1,3 and 5.

■

■ **Example 1.5**

{Fred, dog, 1.45} is the set consisting the objects Fred, dog and 1.45.

■

■ **Example 1.6** If $S=\{x \mid x \text{ a real number}, 0 \leqslant x \leqslant 1\}$, then $1 \in S$ but Fred $\notin S$.

■

Ⓡ Two sets are defined to be equal when they consist of exactly the same elements.

■ **Example 1.7**

$\{1,3,5\}=\{3,5,1\}=\{5,1,3\}$.

■

■ **Example 1.8**

$\{x \mid x^2-1=0\}=\{-1,1\}$.

■

Definition 1.1.3—Subset. The set B is called a **subset** of the set A if and only if $x \in B$ implies $x \in A$.

Notation 1.2 To indicate that B is a subset of A, we write $B \subseteq A$ and say "B is contained in A" or "A contains B".

Definition 1.1.4—Proper subset. If the set B is a subset of the set A, but B is not equal to A, then B is also a **proper** (or **strict**) **subset** of A.

Notation 1.3 To indicate that B is a proper subset of A, we write $B \subset A$.

■ **Example 1.9** If $S=\{\text{pig}, \{\text{dog}\}, \text{cat}\}$, then $\{\text{cat}\} \subset S$, $\{\text{pig}, \{\text{dog}\}\} \subset S$, $\{\{\text{dog}\}\} \subset S$.

■

Definition 1.1.5—Intersection. The set of elements common to two sets A and B is called the **intersection** of A and B (see Figure 1.1, called a **Venn diagram**).

Notation 1.4 To indicate the intersection of A and B, we write $A \cap B$. Consequently, $x \in A \cap B$ if and only if $x \in A$ and $x \in B$.

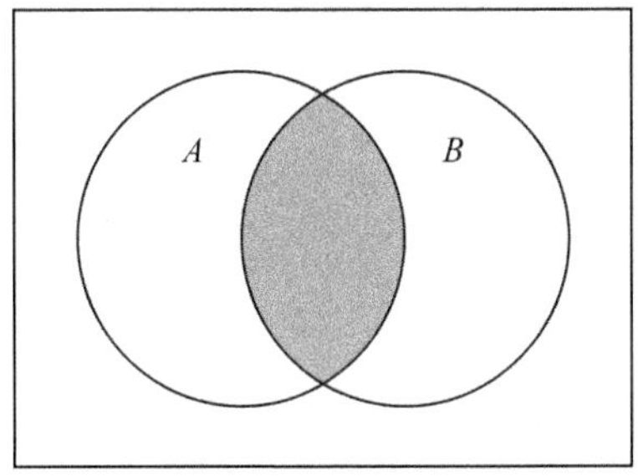

Figure 1.1 The intersection of two sets (shaded region).

■ **Example 1.10** Let A be the set $\{2,3,4,7\}$ and B be the set $\{3,4,5,6,7\}$

$$A \cap B=\{3,4,7\}.$$

■

■ **Example 1.11** Let

$$A=\{\text{pentagon, diamond, square, rectangle}\}$$

and

$$B=\{\text{five pointed star, square, triangle, pentagon}\}.$$

Then

$$A \cap B=\{\text{pentagon, square}\} \text{ (see Figure 1.2).}$$

■

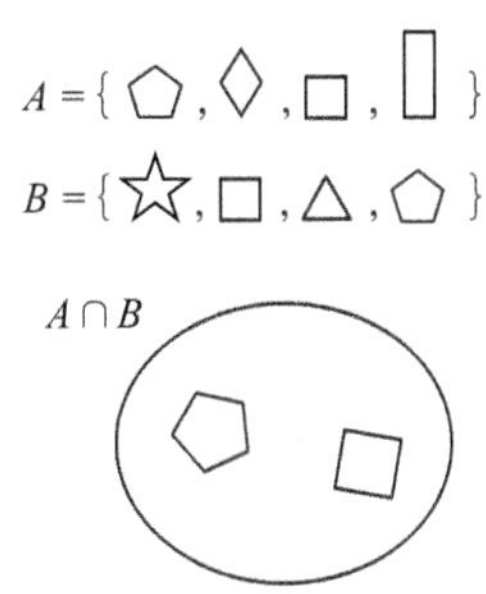

Figure 1.2 The intrtsection of A and B.

Definition 1.1.6—Union. The **union** of two sets A and B is the set of elements that are either in A or in B (see Figure 1.3).

Notation 1.5 To indicate the union of A and B, we write $A \cap B$.

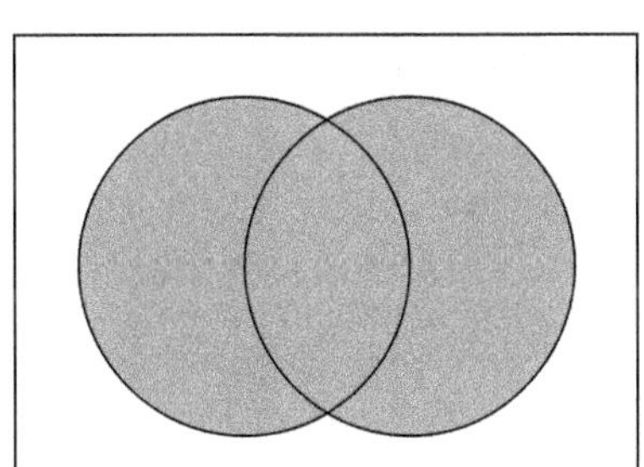

Figure 1.3 The union of two sets (shaded region).

This does not exclude objects that are elements of both A and B.

■ **Example 1.12** Let A be the set $\{2,3,4,7\}$ and B be the set $\{3,4,5,6,7\}$. Then

$$A \cup B = \{2,3,4,5,6,7\}.$$

■

■ **Example 1.13** Let A be the set $\{x \mid x < 1\}$ and B be the set $\{x \mid x \geqslant 0\}$. Then

$$A \cup B = \{x \mid -\infty < x < +\infty\}.$$

■

Definition 1.1.7—Empty set. As a convention, we define the **empty set** to be set consisting of no elements at all.

Notation 1.6 The empty set is denoted by $\varnothing$. If the sets A and B have no elements in common, we say A and B are **disjoint** and write $A \cap B = \varnothing$.

Ⓡ By convention, $\varnothing$ is a subset of every set.

■ **Example 1.14** $1 \notin \varnothing$, since $\varnothing$ has no elements.

■

■ **Example 1.15** Let A be the set $\{1,2,4,7\}$ and B be the set $\{3,5,8,9\}$. Then

$$A \cap B = \varnothing.$$

■

■ **Example 1.16** Let A be the set $\{x \mid x < 1\}$ and B be the set $\{x \mid x \geqslant 0\}$. Then

$$A \cap B = \{x \mid 0 \leqslant x < 1\}.$$

■

■ **Example 1.17** The subsets of $\{1,2\}$ are $\{1,2\}$, $\{1\}$, $\{2\}$ and $\varnothing$.

■

Ⓡ The number of subsets of a given set A is 2^n, in which n is the number of elements of A.

Definition 1.1.8—Complement. Inside a universe U one may define the **complement** of A, also termed the **absolute complement** of A, to be the set of all elements of U not in A.

Notation 1.7 The complement of A is denoted by A^C or $\overline{A}$.

Ⓡ $A \cup A^C = U$.

Definition 1.1.9—Relative complement. The **relative complement** of A with respect to a set B, also termed the difference of sets B and A is the set of elements in B but not in A.

Notation 1.8 The relative complement of A with respect to B is denoted by $B \backslash A$ or $B - A$ or $B \cap A^C$.

■ **Example 1.18** Let A be the set $\{1,2,3,7\}$ and B be the set $\{3,4,5,6,7\}$. Then

$$A \backslash B = \{1,2\}$$

and

$$B \backslash A = \{4,5,6\}.$$

■

Definition 1.1.10—De Morgan's laws. (1) The complement of the union of two sets is the same as the intersection of their complements.

(2) The complement of the intersection of two sets is the same as the union of their complements. See Figure 1.4 for a representation of De Morgan's laws using Venn diagrams.

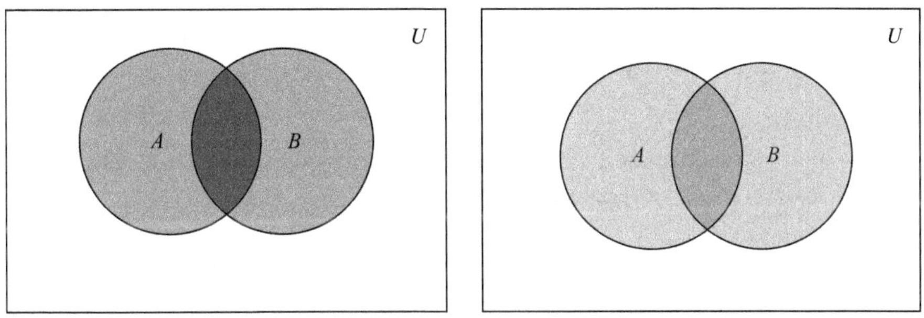

(a) $\overline{A \cup B} = \overline{A} \cap \overline{B}$ (b) $\overline{A \cap B} = \overline{A} \cup \overline{B}$

Figure 1.4 De Morgan's law represented with Venn diagrams.

Notation 1.9 In set theory and Boolean algebra, De Morgan's laws are written formally as

$$\overline{A \cup B} = \overline{A} \cap \overline{B}$$

and

$$\overline{A \cap B} = \overline{A} \cup \overline{B}.$$

Review of the usual sets of numbers

You are already familiar with the following sets of numbers:

Natural numbers. The set of natural numbers is $\{0, 1, 2, 3, 4, \cdots\}$ and is denoted by the symbol **N**. They are also called counting numbers since we use them for counting. Note that the set of natural numbers is an infinite set.

Negative integers. The set of negative integers is $\{\cdots, -6, -5, -4, -3, -2, -1\}$ and is denoted by the symbol **Z**. This is also an infinite set.

Integers. The set of integers is $\{\cdots, -3, -2, -1, 0, 1, 2, 3, \cdots\}$ and is denoted by the symbol **Z**, which is for the German "Zahlen", meaning numbers. It can be seen that the set of integers is made up of the set of positive and negative whole numbers. Remember that this set include zero. One must note that the sets of natural numbers, and negative numbers are subsets of the set of integers.

1.1.2 Rational and irrational numbers

Rational numbers are numbers which can always be written as a **common fraction**, i.e., as an integer divided by another integer. In particular, they are numbers which can be expressed in the form $\frac{a}{b}$, where a and b are integers and b is not zero. The top number, a, is called the **numerator** and bottom number, b, is called the **denominator.**

Notation 1.10 The set of rational numbers is represented by the symbol **Q**, which stands for "quotient".

■ **Example 1.19** $\frac{1}{3}$, $\frac{-2}{7}$, $\frac{27}{5}$ and $\frac{-9}{7}$ are rational numbers.

■

Simple fractions can be positive or negative, proper or improper. In general, a common fraction is said to be a **proper fraction** if the absolute value of the fraction is strictly less than one, that is, if the fraction is greater than -1 and less than 1. A common fraction is said to be an **improper fraction**, or sometimes **top-heavy fraction**, if the absolute value of the fraction is greater than or equal to 1.

■ **Example 1.20** $\frac{1}{3}$, $\frac{-2}{5}$ and $\frac{3}{7}$ are proper fractions, while $\frac{10}{3}$, $\frac{-12}{5}$ and $\frac{53}{2}$ are improper fractions.

■

The numbers that may be represented in the decimal system are the **decimal fractions**, sometimes called **decimal numbers**, or simply **decimals.** Those are rational numbers that may be expressed as a fraction, the denominator of which is a power of ten.

■ **Example 1.21** 0.6, -0.157 and 8.976 are decimal numbers, since

$$0.6=\frac{6}{10},\quad -0.157=-\frac{157}{1\,000},\quad 8.976=\frac{8\,976}{1\,000}.$$

■

The decimal system has been extended to infinite decimals, to represent any real number by using an infinite sequence of digits after the decimal separator. In this context, the usual decimals are sometimes called **terminating decimals.** A **repeating decimal**, sometimes called a **recurring decimal**, is an infinite decimal that, after some place, repeats indefinitely the same sequence of digits. For example, $0.080\,808\,080\,8\cdots = 0.\dot{0}\dot{8}$ (or, in another notation, 0.(08)). An infinite decimal represents a rational number if and only if it is a repeating decimal.

In mathematics, the **irrational numbers** are all real numbers which are not rational numbers.

Notation 1.11 The set of irrational numbers is represented by the symbol **H**, or sometimes by **R\Q.**

■ **Example 1.22** $\sqrt{2}$ is an irrational number. Another well known irrational number is π. You cannot find an exact value for $\pi=3.141\,592\,653\,598\cdots$. Also, the square root of a prime number is an irrational number. Examples include $\sqrt{2}$, $\sqrt{3}$, $\sqrt{5}$, $\sqrt{7}$, $\sqrt{11}$ and $\sqrt{13}$. Also, any rational multiple of an irrational number is an irrational number. For example, $3\sqrt{2}$, $5\sqrt{3}$, $17\sqrt{5}$ are all irrational numbers.

■

Ⓡ The mere presence of a square root or radical sign does not mean that the number is irrational. For example $\sqrt{4}$, $\sqrt{9}$ and $\sqrt{36}$ are all rational numbers. Also, the product of irrational numbers is not necessary irrational, even if those numbers are different from each other. For instance, $\sqrt{2}$ and $2\sqrt{2}$ are irrational, but their product is 4, which is rational.

Exercise 1.1 Sort out the following numbers into rational and irrational, giving reasons.

(1) 0.325;

(2) 0.666 6 …;

(3) 0.090 909 …;

(4) 0.732 334 345 5 ….

■

Solution

(1) 0.325	Rational	(terminating decimal number).
(2) 0.666 6 …	Rational	(non-terminating recurring decimal number).
(3) 0.090 909 …	Rational	(non-terminating recurring decimal number).
(4) 0.732 334 345 5 …	Irrational	(non-terminating non-recurring decimal).

■

Exercise 1.2 Which of the following sets contain rational numbers only?

(1) $P=\{2, \sqrt{3}, 7\}$;

(2) $Q=\{15, \frac{1}{6}, \sqrt{36}\}$;

(3) $R=\{x \mid -3\leqslant x\leqslant 8\}$.

■

Solution

(1) The set P is not a set of rational numbers since $\sqrt{3}$ is an irrational number.

(2) The set Q is a set of rational numbers because $\frac{1}{6}$ is a common fraction and $\sqrt{36}=6$.

(3) Since x can be either rational or irrational in the given interval, the set R is a set of real numbers, and not a set of rational numbers.

■

Operations with rational numbers

Addition and subtraction of fractions

Adding and subtracting fractions are easy when the denominators are the same. When fractions have different denominators, there are a few steps you have to follow to find a common denominator so the fractions can be added to or subtracted from each other.

Ⓡ In mathematics, the lowest common denominator or least common denominator (abbreviated LCD) is the least common multiple of the denominators of a set of fractions. It simplifies adding, subtracting, and comparing fractions.

Exercise 1.3 Perform the following operations:

(1) $\frac{1}{2}+\frac{2}{5}$;

(2) $3\frac{1}{2}+2\frac{2}{3}$;

(3) $\frac{3}{4}-\frac{1}{3}$;

(4) $5\frac{1}{3}-1\frac{11}{12}$.

■

Solution

(1) $\frac{1}{2}+\frac{2}{5}=\frac{5+4}{10}$ (the LCD of 2 and 5 is 10)

$=\frac{9}{10}$.

(2) $3\frac{1}{2}+2\frac{2}{3}=\frac{7}{2}+\frac{8}{3}$ (changing mixed fractions into improper fractions)

$=\frac{21+16}{6}$ (the LCD of 2 and 3 is 6)

$=\frac{37}{6}=6\frac{1}{6}$.

(3) $\frac{3}{4}-\frac{1}{3}=\frac{9-4}{12}$ (the LCD of 4 and 3 is 12)

$=\frac{5}{12}$.

(4) $5\frac{1}{3}-1\frac{11}{12}=\frac{16}{3}-\frac{23}{12}$ (changing mixed fractions into improper fractions)

$=\frac{64-23}{12}$ (the LCD of 3 and 12 is 12)

$=\frac{41}{12}=3\frac{5}{12}$.

■

Multiplication of fractions

To multiply two fractions, multiply top numbers together and bottom numbers together, and simplify if needed. Mixed numbers must first be changed to improper fractions before multiplying.

Exercise 1.4 Perform the following operations:

(1) $\frac{1}{3}\times\frac{4}{5}$;

(2) $2\frac{1}{3}\times1\frac{1}{2}$;

(3) $\frac{14}{15}\times\frac{5}{21}$;

(4) $3\frac{1}{7}\times4\frac{1}{2}\times1\frac{5}{9}$;

(5) $2\frac{1}{2}\times3\frac{1}{3}\times1\frac{5}{6}$;

(6) $\frac{5}{6}\times4\frac{1}{4}\times\frac{8}{17}$. ■

Solution

(1) $\frac{1}{3}\times\frac{4}{5}=\frac{4}{15}$.

(2) $2\frac{1}{3}\times1\frac{1}{2}=\frac{7}{3}\times\frac{3}{2}=\frac{7}{2}=3\frac{1}{2}$.

(3) $\frac{14}{15}\times\frac{5}{21}=\frac{14}{63}=\frac{2}{9}$.

(4) $3\frac{1}{7}\times4\frac{1}{2}\times1\frac{5}{9}=\frac{22}{7}\times\frac{9}{2}\times\frac{14}{9}=\frac{22}{1}=22$.

(5) $2\frac{1}{2}\times3\frac{1}{3}\times1\frac{5}{6}=\frac{5}{2}\times\frac{10}{3}\times\frac{11}{6}=\frac{275}{18}$.

(6) $\frac{5}{6}\times4\frac{1}{4}\times\frac{8}{17}=\frac{5}{6}\times\frac{17}{4}\times\frac{8}{17}=\frac{5}{3}$.

■

Division of fractions

To divide one fraction by another, invert the divisor and then proceed as in multiplication. Mixed numbers must first be changed to improper fractions before dividing.

■ **Example 1. 23** In order to divide $\frac{1}{3}$ by $\frac{2}{5}$, the second fraction is turned upside down to give $\frac{5}{2}$ and then the first fraction, $\frac{1}{3}$, is multiplied by the second fraction turned upside down, $\frac{5}{2}$. The answer is $\frac{5}{6}$. ■

■ **Example 1.24** Divide $\frac{36d}{5}$ by $\frac{18h}{10y}$. The answer is $\frac{36d}{5}\cdot\frac{10y}{18h}=\frac{4dy}{h}$. ■

Exercise 1.5 Perform the following operations:

(1) $\frac{1}{2}\div\frac{1}{4}$;

(2) $3\frac{1}{4}\div 2\frac{1}{3}$;

(3) $\frac{7}{12}\div 1\frac{1}{2}$;

(4) $1\frac{1}{3}\div\frac{1}{4}$;

(5) $\frac{5}{6}\div\frac{1}{4}$;

(6) $\frac{2}{5}\div\frac{5}{2}$. ■

Solution

(1) $\frac{1}{2}\div\frac{1}{4}=\frac{1}{2}\times\frac{4}{1}=2$.

(2) $3\frac{1}{4}\div 2\frac{1}{3}=\frac{13}{4}\div\frac{7}{3}=\frac{13}{4}\times\frac{3}{7}=\frac{39}{28}=1\frac{11}{28}$.

(3) $\frac{7}{12}\div 1\frac{1}{2}=\frac{7}{12}\div\frac{3}{2}=\frac{7}{12}\times\frac{2}{3}=\frac{7}{18}$.

(4) $1\frac{1}{3}\div\frac{1}{4}=\frac{4}{3}\div\frac{1}{4}=\frac{4}{3}\times\frac{4}{1}=\frac{16}{3}=5\frac{1}{3}$.

(5) $\frac{5}{6}\div\frac{1}{4}=\frac{5}{6}\times\frac{4}{1}=\frac{10}{3}=3\frac{1}{3}$.

(6) $\frac{2}{5}\div\frac{5}{2}=\frac{2}{5}\times\frac{2}{5}=\frac{4}{25}$.

■

Ⓡ **BODMAS** rule:

B→Deal with **B**rackets first (parentheses);

O→Deal with **O**f(orders i.e. Powers and Square Roots, Cube Roots, etc.);

DM→Deal with **D**ivision and **M**ultiplication (start from left to right since they perform equally);

AS→Deal with **A**ddition and **S**ubtraction (start from left to right since they perform equally).

Exercise 1.6 Perform the following operations:

(1) $1\frac{2}{3}-\left(1\frac{3}{4}\div 2\frac{5}{8}\right)$;

(2) $\left(1\frac{1}{3}\times 2\frac{1}{4}\right)+\left(2\frac{1}{2}\div\frac{5}{6}\right)$;

(3) $2\frac{1}{3}+\frac{1}{2}\times 3\frac{1}{3}$. ■

Solution

(1) $1\frac{2}{3}-\left(1\frac{3}{4}\div 2\frac{5}{8}\right)=\frac{5}{3}-\left(\frac{7}{4}\div\frac{21}{8}\right)$ (converting mixed fractions to improper)

$=\frac{5}{3}-\left(\frac{7}{4}\times\frac{8}{21}\right)$ (solving bracket first)

$=\frac{5}{3}-\frac{2}{3}=\frac{3}{3}=1.$

(2) $\left(1\frac{1}{3}\times 2\frac{1}{4}\right)+\left(2\frac{1}{2}\div\frac{5}{6}\right)=\left(\frac{4}{3}\times\frac{9}{4}\right)+\left(\frac{5}{2}\div\frac{5}{6}\right)$

$=3+\left(\frac{5}{2}\times\frac{6}{5}\right)$

$=3+3=6.$

(3) $2\frac{1}{3}+\frac{1}{2}\times 3\frac{1}{3}=\frac{7}{3}+\left(\frac{1}{2}\times\frac{10}{3}\right)$ (dealing with multiplication first)

$=\frac{7}{3}+\frac{5}{3}=\frac{12}{3}=4.$

■

Comparing and ordering rational numbers

In general, the procedure of comparing numbers is as follows:

Step 1 Express the given rational numbers as common fractions with positive denominators.

Step 2 Express the rational numbers as equivalent rational numbers with common denominators.

Step 3 Compare the numerators. The rational number with larger numerator is bigger.

■ **Example 1.25** Compare the following rational numbers:

(1) $\frac{4}{7}$ and $\frac{6}{7}$; (2) $\frac{5}{8}$ and $-\frac{6}{8}$; (3) $-\frac{5}{4}$ and $-\frac{7}{4}$.

The answers are:

(1) Both rational numbers have positive denominators and the denominators are common, which implies that $\frac{6}{7}>\frac{4}{7}$ since $6>4$.

(2) $\frac{5}{6}$ is positive while $-\frac{6}{8}$ is negative. Therefore, $\frac{5}{8}>-\frac{6}{8}$.

(3) Both rational numbers have positive denominators and the denominators are common, which implies that $-\frac{5}{4}>-\frac{7}{4}$ since $-5>-7$.

■

Exercise 1.7 Arrange 0.309, 0.299, 0.0999, 0.3 and 0.31 in increasing order. ■

Solution Change the decimals to fractions with common denominator (the LCD of the denominators of the fractions is 10 000) and compare them.

$$0.0999=\frac{999}{10\,000},$$

$$0.299=\frac{299}{1\,000}=\frac{2\,990}{10\,000},$$

$$0.3=\frac{3}{10}=\frac{3\,000}{10\,000},$$

$$0.309=\frac{309}{1\,000}=\frac{3\,090}{10\,000},$$

$$0.31=\frac{31}{100}=\frac{3\,100}{10\,000}.$$

Therefore,

$$\frac{999}{10\,000}<\frac{2\,990}{10\,000}<\frac{3\,000}{10\,000}<\frac{3\,090}{10\,000}<\frac{3\,100}{10\,000},$$

that is

$$0.0999<0.299<0.3<0.309<0.31.$$

■

Ⓡ When the rational numbers to be compared are given in different forms, first write all of them in the same form.

■ **Example 1.26** Mr. Plum's science class is growing plants under different conditions. The average plant growth during a week was 5.5 inches. Figure 1.5 shows the difference (in inches) from the average for some students' plants. Order the differences from least to greatest. Express each number as a decimal.

Ricky's plant: $3\frac{1}{4}=3.25$ Debbie's plant: -2.2

Suni's plant: 1.7 Leonora's plant: $-1\frac{7}{10}=-1.7$

From least to greatest, the difference are -2.2, $-1\frac{7}{10}$, 1.7 and $3\frac{1}{4}$.

■

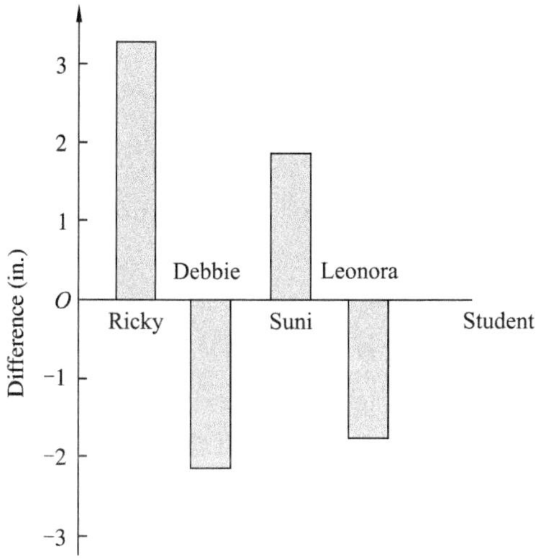

Difference from Average Plant Growth	
Student	Difference (in.)
Ricky	$3\frac{1}{4}$
Debbie	-2.2
Suni	1.7
Leonora	$-1\frac{7}{10}$

Figure 1.5 Difference from the average for some students' plants.

Rounding off decimals

To correct a decimal to a stated number of decimal places:

(1) Find the place value you want (the "rounding digit") and look at the digit just to the right of it.

(2) If that digit is less than 5, do not change the rounding digit but drop all digits to the right of it.

(3) If that digit is greater than or equal to 5, add one to the rounding digit and drop all digits to the right of it.

■ **Example 1.27** Correct 23.462 5 to

(1) 1st decimal place; (2) 2nd decimal place; (3) 3rd decimal place.

The answers are:

(1) The 1st d.p. is 4 but the figure after it is 6 which is more than 5. Therefore we add 1 to 4 to give 5 in the 1st d.p. such that $23.4625\approx 23.5$ (correct to 1st d.p.).

(2) The 2nd d.p. is 6 but the figure after it is 2 which is less than 5. We then leave 6 as it is. Therefore, $23.4625\approx 23.46$ (correct to 2nd d.p.).

(3) The 3rd d.p. is 2 but the figure after it is 5. We then add one to 2 to give 3 in the 3rd d.p. Therefore, $23.4625\approx 23.463$ (correct to 3rd d.p.).

■

Rounding numbers to the nearest tenth, hundredth, thousandth etc.

For any number with a decimal part, the first decimal place is called the tenths, the second decimal place is called the hundredths, the third decimal place is called thousandths, etc.

For example, 987.345 6 rounded to the nearest

(1) whole number is 987;

(2) tenth is (i.e. 1st d.p.) 987.3;

(3) hundredth is (i.e. 2nd d.p.) 987.35;

(4) thousandth is (i.e. 3rd d.p.) 987.346.

Decimal approximation

The place of a figure in a number is important. It tells you what the figure is worth. As you read a number from left to right, the first value you come to that is not zero has the highest place value, so it is called the first significant figure (s.f.). For example, in the number 306, 3 is the first significant figure, 0 is the second and 6 is the third. Also, in the number 0.000 406, 4 is the first significant figure since it is the first non-zero figure reading from left to right. The zero after 4 is the second significant figure and 6 is the third. To correct a number to a stated number of significant figures, you may proceed as follows:

(1) Find the last significant figure you want.

(2) Then look at the next significant figure (to the right) if this figure is less than 5 leave the last significant figure you want as it is. If this figure is 5 or more add 1 to the last significant figure you want.

Ⓡ **Significant figure rules:**

Rule 1 Non-zero digits are always significant.

Rule 2 Any zeros between two significant digits are significant.

Rule 3 A final zero or trailing zeros in the decimal portion are significant.

Exercise 1.8

(1) Correct 0.009 416 to 1 significant figure.

(2) Correct 15.072 to 3 significant figures.

(3) Round 287530 to 4 significant figures.

■

Solution

(1) 0.009. (to 1 significant figure)

(2) 15.1. (to 3 significant figures)

(3) 287 500. (to 4 significant figures)

■

Expressing very large or very small numbers in standard form

Sometimes we come across very large numbers or very small numbers in our daily lives. Since it becomes very cumbersome to express such numbers in their full form (picture yourself, for instance, trying to read a number consisting by a 3 followed by 25 zeroes), standard form is used to make them manageable. In standard form, a positive number is written as

$$a\times 10^n \text{ with } 1\leqslant a<10 \text{ and } n \text{ an integer},$$

in which the value of n in standard form shows whether the number is greater than 1 or not. This form is often much easier to write, read and grasp (for instance, the previously alluded number can just be written as 3×10^{25}).

To change a number to standard form

First, write down the number from 1 and less than 10 and then work out the power of 10. To work out the number from 1 and less than 10, put the decimal point between the first and second significant figures of the number. For example, for 345 the number part is 3.45 and for 0.000 423 the number part is 4.23.

To work out the power of 10.

(1) Compare the ordinary number to the number part of its standard form.

(2) Count the number of places you have to move the number part from the decimal point to get the ordinary number. This tells you the power 10 to multiply.

Exercise 1.9 Write the following in standard form:

(1) 2 000;

(2) 41 460;

(3) 4 687.02;

(4) 3 201.7;

(5) 0.050 03;

(6) 0.000 036 27.

■

Solution

(1) $2\,000=2.0\times10^{3}$.

(2) $41\,460=4.146\times10^{4}$.

(3) $4\,687.02=4.687\,02\times10^{3}$.

(4) $3\,201.7=3.201\,7\times10^{3}$.

(5) $0.050\,03=5.003\times10^{-2}$.

(6) $0.000\,036\,27=3.627\times10^{-5}$.

■

1.1.3 Real numbers

In mathematics, a **real number** is a value that represents a quantity along a line. The adjective **real** in this context was introduced in the 17th century by René Descartes. The real numbers include all the rational numbers and all the irrational numbers. Included within the irrationals are the transcendental numbers. Real numbers can be thought of as points on an infinitely long line called the **number line** or **real line**, where the points corresponding to integers are equally spaced (see Figure 1.6).

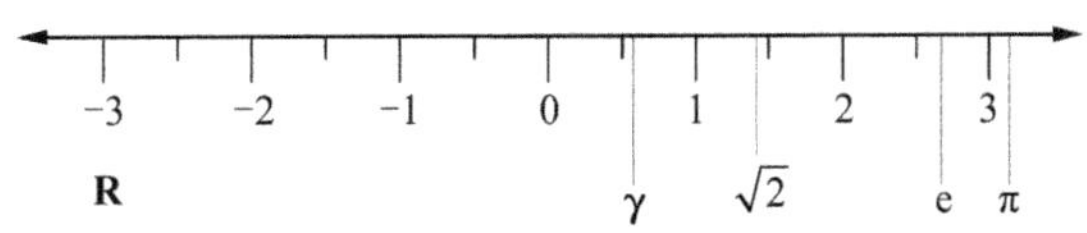

Figure 1.6 Real numbers can be thought of as points on an infinitely long number line.

Notation 1.12 The set of real numbers is represented by the symbol **R**.

® **R**= **Q**∪**H** and **N**⊂**Z**⊂**Q**⊂**R**.

Inequality symbols

Symbol	Read
$a>b$	a is greater than b
$a<b$	a is less than b
$a\geqslant b$	a is greater than or equal to b
$a\leqslant b$	a is less than or equal to b

Bounded intervals

Inequality notation	Interval notation	Interval type
$a\leqslant x\leqslant b$	$[a,b]$	Closed
$a<x<b$	(a,b)	Open
$a\leqslant x<b$	$[a,b)$	Half-open
$a<x\leqslant b$	$(a,b]$	Half-open

Unbounded intervals

Inequality notation	Interval notation	Interval type
$x\geqslant a$	$[a, +\infty)$	Closed
$x>a$	$(a,+\infty)$	Open
$x\leqslant b$	$(-\infty,b]$	Closed
$x<b$	$(-\infty,b)$	Open

■ **Example 1.28** Write an interval of real numbers using inequality notation and interval notation:

(1) the real numbers between -4 and -0.5;

(2) the real numbers greater than or equal to zero.

The answers are:

(1) inequality notation: $-4\leqslant x\leqslant -0.5$ and interval notation: $[-4,-0.5]$;

(2) inequality notation: $x\geqslant 0$ and interval notation: $[0,+\infty)$.

■

■ **Example 1.29** State whether the following intervals are bounded or not and their types:

(1) $[-6,3)$;

(2) $(-\infty,0)$;

(3) $-1\leqslant x\leqslant 3$.

The answers are:

(1) it is bounded and half-open;

(2) it is unbounded and open;

(3) it is bounded and closed.

■

Properties of algebra

Commutative properties

$a+b=b+a$,

$ab=ba$.

Associative properties

$(a+b)+c=a+(b+c)$,

$(ab)c=a(bc)$.

Distributive properties

$a(b+c)=ab+ac$,

$(b+c)a=ab+ac$.

Identity properties

$u+0=u$,

$u\cdot 1=u$.

Inverse properties

$u+(-u)=0$,

$u\cdot \frac{1}{u}=1$.

■ **Example 1.30** State the property of algebra being used.

(1) $7+10=10+7$. (commutative property)

(2) $(2+3)+4=2+(3+4)$. (associative property)

(3) $(x+2y)+3z=x+(2y+3z)$. (associative property)

(4) $3(A+B)=3A+3B$. (distributive property)

■

1.2 Indices

The index form or index notation is a simple way for writing numbers. For example, the number 3^2 is in the index form. The small 2 is the index and the 3 is called the base. In general:

a^5 means $a\times a\times a\times a\times a$ [5 factors],

a^n means $a\times a\times\cdots\times a\times a$ [n factors],

where n is the **index** (plural indices) and a is the base. The index is also called the **power** or the **exponent**. An index can be positive or negative or a fraction.

1.2.1 Laws of indices

All numbers written with indices obey the same rules.

First law of indices

We know that

$$a^2 \times a^3 = (a \times a) \times (a \times a \times a) = a^{2+3} = a^5.$$

This specifically gives the first law indices, also known as the **product law**.

$$a^m \times a^n = a^{m+n}.$$

■ **Example 1.31**

$$2^2 \times 2^3 = 2^{2+3} = 2^5.$$

$$(2x^4 y^{-3})(4x^{-2} y^6) = 8x^{4+(-2)} y^{-3+6} = 8x^2 y^3.$$

$$x^2 \times x^{-3} = x^{2-3} = x^{-1}.$$

■

■ **Example 1.32**

$(a+b)^2 = (a+b)(a+b) = a^2 + 2ab + b^2.$

$(a-b)^2 = (a-b)(a-b) = a^2 - 2ab + b^2.$

$(a+b)^3 = (a+b)^2(a+b) = (a^2 + 2ab + b^2)(a+b) = a^3 + 3a^2 b + 3ab^2 + b^3.$

$(a-b)^3 = (a-b)^2(a-b) = (a^2 - 2ab + b^2)(a-b) = a^3 - 3a^2 b + 3ab^2 - b^3.$

$(a+b+c)^2 = (a+b+c)(a+b+c) = a^2 + b^2 + c^2 + 2ab + 2ac + 2bc.$

■

Second law of indices

The second law of indices is also known as the **quotient law**.

We know that

$$\frac{a^5}{a^3} = \frac{a \times a \times a \times a \times a}{a \times a \times a} = a^{5-3} = a^2.$$

In general,

$$a^m \div a^n = a^{m-n} \text{ or } \frac{a^m}{a^n} = a^{m-n}.$$

■ **Example 1.33**

$$2^4 \div 2^2 = 2^{4-2} = 2^2 = 4.$$

$$3^6 \div 3^3 = 3^3 = 27.$$

$$\frac{x^a}{x^b} = x^{a-b}.$$

■

Third law of indices

The third law of indices is also known as the **power law**. For example,

$$(a^3)^2 = (a^3 \times a^3) = (a \times a \times a) \times (a \times a \times a) = a^{3+3} = a^6.$$

In general,

$$(a^m)^n = a^{mn}.$$

■ **Example 1.34**

$$(2^4)^5 = 2^{4 \times 5} = 2^{20}.$$

$$(x^2)^4 = x^8.$$

$$(a^2 b^3)^2 = a^4 b^6.$$

■

1.2.2 Properties of indices

The following properties of indices should be noted carefully in addition to the three basic laws above. They are used to simplify expressions.

Property 1: Power zero

$$a^0=1,$$

that is, any non-zero number raised to the power zero is equal to 1. For example,

$$7^0=1,\ 100^0=1,\ 120^0=1.$$

Property 2: Power of product of two numbers

$$(ab)^n=a^nb^n,$$

that is, the power of the product is equal to the product of the powers. For example,

$$(2\times3)^2=2^2\times3^2=4\times9=36.$$

Property 3: Negative index

$$a^{-n}=\frac{1}{a^n},\ a\neq0,$$

that is, any number raised to a negative index (power) is equal to the reciprocal of the number raised to the corresponding positive index. For example,

$$3^{-4}=\frac{1}{3^4}=\frac{1}{81}.$$

Ⓡ Any number raised to the power -1 is equal to the reciprocal of the number. For example, $a^{-1}=\frac{1}{a}$.

Property 4: Power of quotient of two numbers

$$\left(\frac{a}{b}\right)^n=\frac{a^n}{b^n},\ b\neq0,$$

that is, the power of the quotient is equal to the quotient of the powers. For example,

$$\left(\frac{2}{3}\right)^2=\frac{2^2}{3^2}=\frac{4}{9}.$$

Property 5: Negative power of quotient of two numbers

$$\left(\frac{a}{b}\right)^{-n}=\left(\frac{b}{a}\right)^n=\frac{b^n}{a^n},\ a\neq0,\ b\neq0,$$

that is, the negative power of a quotient is equal to the corresponding positive power of the quotient turned upside down.

Property 6: Fractional index

(1) $a^{\frac{1}{2}}=\sqrt{a}$.

(2) $a^{\frac{1}{3}}=\sqrt[3]{a}$.

(3) $a^{\frac{1}{4}}=\sqrt[4]{a}$.

In general,

$$a^{\frac{1}{n}}=\sqrt[n]{a},$$

that is, any number raised to the power $\frac{1}{n}$ is equal to the nth root of the number. Also,

$$a^{\frac{m}{n}}=\sqrt[n]{a^m}.$$

Exercise 1.10 Write the following expressions in index form:

(1) $\frac{2}{x^3}$;

(2) $\frac{1}{2x^4}$;

(3) $\sqrt[3]{\frac{54x^4}{2x}}$;

(4) $(x^6)^{\frac{1}{3}}$;

(5) $\left(1\frac{11}{25}\right)^{\frac{1}{2}}$.

■

Solution

(1) $\frac{2}{x^3}=2x^{-3}$.

(2) $\frac{1}{2x^4}=\frac{1}{2}x^{-4}$.

(3) $\sqrt[3]{\frac{54x^4}{2x}}=\sqrt[3]{27x^3}=3x$.

(4) $(x^6)^{\frac{1}{3}}=x^2$.

(5) $\left(1\frac{11}{25}\right)^{\frac{1}{2}}=\left(\frac{36}{25}\right)^{\frac{1}{2}}=\frac{6}{5}$.

■

■ **Example 1.35** Without using mathematical tables, simplify the following:

(1) $\left(\frac{16}{81}\right)^{-\frac{3}{4}}=\left(\frac{81^{\frac{3}{4}}}{16}\right)=\left(\frac{3^4}{2^4}\right)^{\frac{3}{4}}=\left(\frac{3^{4\times\frac{3}{4}}}{2^{4\times\frac{3}{4}}}\right)=\frac{3^3}{2^3}=\frac{27}{8}$;

(2) $\left(\frac{64}{25}\right)^{-\frac{1}{2}}=\left(\frac{25}{64}\right)^{\frac{1}{2}}=\left(\frac{5^2}{8^2}\right)^{\frac{1}{2}}=\left(\frac{5^{2\times\frac{1}{2}}}{8^{2\times\frac{1}{2}}}\right)=\frac{5}{8}$;

(3) $\left(\frac{8}{27}\right)^{-\frac{2}{3}}=\left(\frac{27}{8}\right)^{\frac{2}{3}}=\left(\frac{3^3}{2^3}\right)^{\frac{2}{3}}=\left(\frac{3^{2\times\frac{2}{3}}}{2^{3\times\frac{2}{3}}}\right)=\frac{3^2}{2^2}=\frac{9}{4}$.

■

Exercise 1.11 Simplify the following:

(1) $(27\times3^{-2})(8\times2^{-3})$;

(2) $256^{-\frac{1}{4}}(225^{\frac{1}{2}}+58^0)$;

(3) $27^{\frac{2}{3}}\times64^{\frac{1}{3}}\div81^{\frac{1}{4}}$.

■

Solution

(1) $(27\times3^{-2})(8\times2^{-3})=(3^3\times3^{-2})(2^3\times2^{-3})$

$$=(3^{3+(-2)})(2^{3+(-3)}) \quad \text{(using first law of indices)}$$
$$=3^1\times2^0 \quad \text{(using property 1)}$$
$$=3\times1$$
$$=3.$$

(2) $256^{-\frac{1}{4}}(225^{\frac{1}{2}}+58^0)=\left(\frac{1}{256^{\frac{1}{4}}}\right)[(15^2)^{\frac{1}{2}}+1]$ (using properties 1 and 3)

$$=\frac{1}{\sqrt[4]{256}}[15^{2\times\frac{1}{2}}+1] \quad \text{(using property 6)}$$
$$=\frac{1}{4}(15+1)$$
$$=\frac{1}{4}\times16$$
$$=4.$$

(3) $27^{\frac{2}{3}}\times64^{\frac{1}{3}}\div81^{\frac{1}{4}}=(3^3)^{\frac{2}{3}}\times(4^3)^{\frac{1}{3}}\div(3^4)^{\frac{1}{4}}$

$$=3^2\times4\div3 \quad \text{(using first law of indices)}$$
$$=3\times4$$
$$=12.$$

■

1.2.3 Exponential equations

Exponential equations are equations which have the variable or unknown quantity as an index or exponent. To solve exponential equations without logarithms, you need to have equations with comparable exponential expressions on either side of the "equals" sign, so you can compare the powers and solve. In other words, you have to have "(some base) to (some power) equals (the same base) to (some other power)", then you set the two powers equal to each other, and solve the resulting equation. Such equations can be transformed to linear equations, quadratic equations and simultaneous equations. In this subsection, we only introduce the simplest case and other cases will be introduced in the following sections.

Exponential equations leading to linear equations

To solve such equations we always use the fact that if $a^x=a^y$ implies $x=y$ provided that a is not -1, 0 or 1.

Exercise 1.12 Find the value of x in the following equations:

(1) $3^x=9$;

(2) $3^{x+1}=9^4$;

(3) $64^x=16$;

(4) $3^{\sqrt{x}}=9$;

(5) $8^{x-2}=2^{x-8}$.

■

Solution Note that we write the equation in such a way that the left hand side (L.H.S.) and the right hand side (R.H.S.) both have a common base so that we can equate the exponents.

(1) $3^x=9$,

$3^x=3^2$, (L.H.S. =R.H.S.)

$x=2$. (equating the exponents)

(2) $3^{x+1}=9^4$,

$3^{x+1}=(3^2)^4$,

$3^{x+1}=3^8$, (L.H.S. =R.H.S.)

$x+1=8$, (equating the exponents)

$x=7$.

(3) $64^x=16$,

$(4^3)^x=4^2$,

$4^{3x}=4^2$, (L.H.S. =R.H.S.)

$3x=2$, (equating the exponents)

$x=\dfrac{2}{3}$.

(4) $3^{\sqrt{x}}=9$,

$3^{\sqrt{x}}=3^2$, (L.H.S. =R.H.S.)

$\sqrt{x}=2$, (equating the exponents)

$x=4$.

(5) $8^{x-2}=2^{x-8}$,

$(2^3)^{x-2}=2^{x-8}$,

$2^{3x-6}=2^{x-8}$, (L.H.S. =R.H.S.)

$3x-6=x-8$, (equating the exponents)

$3x-x=-8+6$, (collecting similar terms)

$2x=-2$,

$x=-1$.

■

Exercise 1.13 Find the value of x in the following equations:

(1) $3^{x-2}=\dfrac{1}{81}$;

(2) $2^{3x-1}=128\sqrt{2}$;

(3) $8^x=0.25$.

■

Solution

(1) $3^{x-2}=\frac{1}{81}$,

$3^{x-2}=81^{-1}$,

$3^{x-2}=3^{-4}$, (L.H.S. =R.H.S.)

$x-2=-4$, (equating the exponents)

$x=-2$.

(2) $2^{3x-1}=128\sqrt{2}$,

$2^{3x-1}=2^{7}\times 2^{\frac{1}{2}}$,

$2^{3x-1}=2^{7\frac{1}{2}}$, (L.H.S. =R.H.S.)

$3x-1=7\frac{1}{2}$, (equating the exponents)

$3x=8\frac{1}{2}$,

$x=\frac{17}{6}$.

(3) $8^{x}=0.25$,

$2^{3x}=\frac{1}{4}$, (changing the decimal to a fraction)

$2^{3x}=2^{-2}$, (L.H.S. =R.H.S.)

$3x=-2$, (equating the exponents)

$x=-\frac{2}{3}$.

1.3 Polynomials

Definition 1.3.1—Polynomial. In mathematics, a **polynomial** is an algebraic expression that can be obtained from constants and symbols called indeterminates or variables by means of addition, multiplication and exponentiation to a non-negative integer power.

Polynomials in a single variable

We shall start with polynomials in a single variable. As seen from the above definition, polynomials in the variable (or indeterminate) x are then algebraic expressions that consist in sums of terms of the form ax^n, where n is a non-negative integer and a is a real number, called the **coefficient** of this term. The **degree** of a polynomial in a single variable is the largest exponent of that polynomial.

Definition 1.3.2—Polynomial function. A **polynomial function** is an expression of the form

$$f(x)=a_nx^n+a_{n-1}x^{n-1}+a_{n-2}x^{n-2}+\cdots+a_1x+a_0,$$

in which n is a non-negative integer, x is a variable and a_n, a_{n-1}, a_{n-2}, $\cdots$, a_1, a_0 are constant coefficients. A shorthand for this is the summation notation:

$$\sum_{k=0}^{n} a_kx^k.$$

That is, a polynomial can either be zero or can be written as the sum of a finite number of non-zero terms. If $a_n\neq 0$, then n is called the **degree** of the polynomial, i.e., the highest power of x. Also, a_nx^n is called the **leading term**, a_n is called the **leading coefficient** and a_0 is called the **constant term.**

■ **Example 1.36** $f(x)=4x^3+4x^2-2x+8$ is a polynomial function of degree 3, with leading term $4x^3$, leading coefficient 4 and constant term 8.

■

■ **Example 1.37** $f(x)=3x^{-2}$ is not a polynomial function because x has a negative exponent, instead of a non-negative one. Also, $g(x)=\sqrt{x}$ is not a polynomial function because polynomials do not contain square roots (or, equivalently, because $g(x)=x^{\frac{1}{2}}$, and therefore x has a fractional exponent, instead of an integer one).

■

Ⓡ By definition, the degree of the zero polynomial is either left explicitly undefined, or is defined to be negative (usually -1 or $-\infty$).

In what follows, we shall limit ourselves to polynomials of degree not exceeding 4.

Particular polynomial functions

Lower degree polynomial functions have special names, as they occur more frequently. The three types of polynomial functions which will be considered at this level are:

(1) **Linear** functions (for example, $f(x)=3x+5$) (degree=1)

(2) **Quadratic** functions (for example, $f(x)=4x^2+3x-5$) (degree=2)

(3) **Cubic** functions (for example, $f(x)=3x^3+2x^2-9x+4$) (degree=3)

Graphing linear and quadratic functions

1. The graph of a **liniar function** is a **straight line.**

2. The graph of a **quadratic function** $f(x)=ax^2+bx+c$ is a **parabola.** If $a>0$, the parabola opens upwards. If $a<0$, the parabola opens downwards. The coefficients b and a together control the position of the axis of symmetry of the parabola, which is $x=-\frac{b}{2a}$. The coefficient c controls the height of the parabola; more specifically, it is the height of the parabola where it intercepts the y-axis. Here, the **vertex** of a parabola is the place where it turns; hence, it is also called the **turning point**, which is given by $\left(-\frac{b}{2a}, c-\frac{b^2}{4a}\right)$.

Polynomials in two variables

Next, let us take a quick look at polynomials in two variables. Polynomials in two variables are algebraic expressions consisting in sums of terms of the form $ax^m y^n$. The degree of each term in a polynomial in two variables is the sum of the exponents in each term and the degree of the polynomial is the largest of those sums. Some examples of polynomials in two variables are:

(1) $f(x,y)=x^2y^2-xy^3+2y^4+56$;

(2) $f(x,y)=x^4+y^3-6x+9y-9$.

In this kind of polynomials, not every term needs to contain both x and y. In fact, as seen in the second example they may not have any term that contains both x and y. Also, the degree of the polynomial may come from terms involving only one variable or multiple terms, as seen from the first example. In both examples, the polynomial functions have the same degree, 4.

Polynomials with fewer terms

"Polynomial" literally means "many parts" (terms). Polynomials with fewer terms have special names as well, with the actual number of terms replacing "poly".

Definition 1.3.3—Monomial, Binomial, Trinomial. A **monomial** is a polynomial that consists of exactly one term. A **binomial** is a polynomial that consists of exactly two terms. A **trinomial** is a polynomial that consists of exactly three terms.

■ **Example 1.38** $f(x,y)=2x^2y^4$ is a monomial in the variables x, y. Also, $g(x)=5x^2-7$ is a binomial in the variable x, while $h(x,y)=2x^4-5xy^2+9y$ is a trinomial in the variables x,y.

■

Operations with polynomials

Addition and subtraction

Polynomials can be added using the associative law of addition and grouping terms with similar degree. The coefficient of a term with a certain degree will then be the sum of the coefficients with the same degree from each polynomial (if a term with that degree is missing from a polynomial, we could consider that the coefficient is 0). A similar rule holds for subtraction. In fact, the addition or subtraction of polynomials reduces to the addition or subtraction of their coefficients.

Exercise 1.14 Find $f(x)+g(x)$ provided that:

(1) $f(x)=3x^3+4x^2-2x+8$ and $g(x)=2x^3+3x^2+5x+2$;

(2) $f(x)=3x^3+4$ and $g(x)=2x^2+3x+5$;

(3) $f(x)=7x^3+x-4$ and $g(x)=4x^3+2x^2-2$.

■

Solution

(1) $f(x)=3x^3+4x^2-2x+8,$

$g(x)=2x^3+3x^2+5x+2,$

$f(x)+g(x)=5x^3+7x^2+3x+10.$

(2) $f(x)=3x^3+0x^2+0x+4,$

$g(x)=0x^3+2x^2+3x+5,$

$f(x)+g(x)=3x^3+2x^2+3x+9.$

(3) $f(x)=7x^3+0x^2+x-4,$

$g(x)=4x^3+2x^2+0x-2,$

$f(x)+g(x)=11x^3+2x^2+x-6.$

■

Exercise 1.15 Find $h(x)-p(x)$ if

(1) $h(x)=5x^3+2x^2+x+5$ and $p(x)=2x^3+4x+8$;

(2) $h(x)=7x^3+x-4$ and $p(x)=4x^3-3x^2+4x+3$.

■

Solution

(1) $h(x)=5x^3+2x^2+x+5,$

$p(x)=2x^3+0x^2+4x+8,$

$h(x)-p(x)=3x^3+2x^2-3x-3.$

(2) $h(x)=7x^3+0x^2+x-4,$

$p(x)=4x^3-3x^2+4x+3,$

$h(x)-p(x)=3x^3+3x^2-3x-7.$

■

Multiplication

The multiplication of polynomials is more complicated than their addition or subtraction. A good mnemonic for the multiplication of two binomials (obtained via the distributive law) is the **FOIL** rule.

FOIL rule:

F→Take the product of **F**irst terms;

O→Take the product of **O**uter terms;

I→Take the product of **I**nner terms;

L→Take the product of **L**ast terms.

$$\begin{aligned}(2x+3)(4x-5)&=2x\times4x+2x\times(-5)+3\times4x+3\times(-5)\\&=8x^2-10x+12x-15\\&=8x^2+2x-15.\end{aligned}$$

To work out the product of two general polynomials into a sum of terms, the distributive law is repeatedly applied, which results in each term of one polynomial being multiplied by every term of the other.

Exercise 1.16 Find $f(x)\times g(x)$ if

(1) $f(x)=2x+4$ and $g(x)=x+3$;

(2) $f(x)=4x^2$ and $g(x)=x^2-6x+2$;

(3) $f(x)=2x^2+4x+2$ and $g(x)=3x^2+5x+4$.

■

Solution

(1) $$\begin{aligned} f(x)\times g(x) &= (2x+4)(x+3) \\ &= 2x(x+3)+4(x+3) \\ &= 2x^2+6x+4x+12 \\ &= 2x^2+10x+12. \end{aligned}$$

(2) $$\begin{aligned} f(x)\times g(x) &= 4x^2(x^2-6x+2) \\ &= 4x^4-24x^3+8x^2. \end{aligned}$$

(3) $$\begin{aligned} f(x)\times g(x) &= (2x^2+4x+2)(3x^2+5x+4) \\ &= 2x^2(3x^2+5x+4)+4x(3x^2+5x+4)+2(3x^2+5x+4) \\ &= 6x^4+22x^3+34x^2+26x+8. \end{aligned}$$

■

Exercise 1.17 Expand and simplify:

(1) $(2x+6)^2$;

(2) $4(x+3)^3$;

(3) $(3x^3-x^2)(2x^3-3x)$.

■

Solution

(1) $$\begin{aligned} (2x+6)^2 &= (2x+6)(2x+6) \\ &= 4x^2+12x+12x+36 \\ &= 4x^2+24x+36. \end{aligned}$$

(2) $$\begin{aligned} 4(x+3)^3 &= 4(x+3)(x+3)(x+3) \\ &= 4(x^2+6x+9)(x+3) \\ &= 4(x^3+6x^2+9x+3x^2+18x+27). \\ &= 4x^3+36x^2+108x+108. \end{aligned}$$

(3) $$\begin{aligned} (3x^3-x^2)(2x^3-3x) &= 3x^3(2x^3-3x)-x^2(2x^3-3x) \\ &= 6x^6-2x^5-9x^4+3x^3. \end{aligned}$$

■

Ⓡ $$(a\pm b)^2 \neq a^2 \pm b^2.$$

1.4 Factorization

Definition 1.4.1—Factorization. In mathematics, **factorization** or **factoring** is the decomposition of a mathematical object (for example, a number or a polynomial) into a product of several **factors**, usually smaller or simpler objects of the same kind.

Factorization can be shown with natural numbers. The integers that divide exactly into 15 are 1, 3, 5 and 15. These are called **factors** of 15. Note that 15 can be written as a product of some of these factors.

$$15=1\times 15,\ 15=3\times 5.$$

Polynomials also can sometimes be factorized. One such quadratic expression is $x^2+3x-18$. It can be factorized as follows

$$x^2+3x-18=(x+6)(x-3).$$

Here we call $(x+6)$ and $(x-3)$ the factors of the quadratic $x^2+3x-18$.

Several basic techniques for factorization are listed below.

Extracting a common factor

The general method used to factorize terms with common factors is to find the factor which is common to all the terms in the expression and bring it outside the bracket. Inside the brackets, one finds the old terms divided by what was taken outside the bracket.

Exercise 1.18 Factorize the following expressions:

(1) $3ax+6ay$;

(2) $54-81x$;

(3) $100x-25x^2$.

■

Solution

(1) $3ax+6ay=3a\left(\frac{3ax}{3a}+\frac{6ay}{3a}\right)=3a(x+2y)$.

(2) $54-81x=27\left(\frac{54}{27}-\frac{81x}{27}\right)=27(2-3x)$.

(3) $100x-25x^2=25x\left(\frac{100x}{25x}-\frac{25x^2}{25x}\right)=25x(4-x)$.

■

Exercise 1.19 Factorize the following expressions:

(1) $6x^3+2x^4+4x^5$;

(2) $7a-7a^3+14a^4$;

(3) $3a^2-6ax$.

■

Solution

(1) $6x^3+2x^4+4x^5=2x^3(3+x+2x^2)$.

(2) $7a-7a^3+14a^4=7a(1-a^2+2a^3)$.

(3) $3a^2-6ax=3a(a-2x)$.

■

Exercise 1.20 Without using a calculator, evaluate $16.5\times6.67-4.67\times16.5$.

■

Solution

$$\begin{aligned}16.5\times6.67-4.67\times16.5&=16.5(6.67-4.67)\\&=16.5\times2\\&=33.\end{aligned}$$

■

A grouping method

An expression may be factored if all terms can be arranged in groups which have a common factor. Many of those expressions have four terms. The general method used to factorize such expressions is to group the first two terms and the last two terms separately. Next, factorize each group on its own. In doing so the terms in the brackets will (hopefully) be the same. Then put the factors in each group inside one more set of brackets and multiply by the common bracket.

Exercise 1.21 Factorize completely $x^2-ax+bx-ab$.

■

Solution Noticing that the first two terms contain a common factor x and for the last two terms a common factor is b, we enclose the first two terms in one bracket and the last two in another. i.e.,

$$(x^2-ax)+(bx-ab).$$

We then factorize each term separately so that the terms in the bracket will be the same. i.e.,

$$x(x-a)+b(x-a).$$

Finally, we pick the terms outside each bracket and put them inside one more set of brackets, while multiplying with the common bracket. Therefore we have

$$\begin{aligned}x^2-ax+bx-ab&=(x^2-ax)+(bx-ab) &&\text{(grouping)}\\&=x(x-a)+b(x-a) &&\text{(factorizing each term separately)}\\&=(x-a)(x+b).\end{aligned}$$

■

Exercise 1.22 Factorize the following expressions completely:

(1) $3a^2+2ab-12ac-8bc$;

(2) $2ap+aq-bq-2bp$;

(3) $mp+np-mt-nt$;

(4) $y(6x-2)-z(3x-1)$.

■

Solution

(1) $3a^2+2ab-12ac-8bc=(3a^2+2ab)-(12ac+8bc)$ (grouping)

$=a(3a+2b)-4c(3a+2b)$ (factorizing)

$=(3a+2b)(a-4c)$.

(2) $2ap+aq-bq-2bp=(2ap+aq)-(bq+2bp)$ (grouping)

$=a(2p+q)-b(q+2p)$ (factorizing)

$=(2p+q)(a-b)$.

(3) $mp+np-mt-nt=(mp+np)-(mt+nt)$ (grouping)

$=p(m+n)-t(m+n)$ (factorizing)

$=(m+n)(p-t)$.

(4) $y(6x-2)-z(3x-1)=2y(3x-1)-z(3x-1)$ (factorizing)

$=(3x-1)(2y-z)$.

■

Factorization of quadratic trinomials

Any expression of the form ax^2+bx+c, where a, b and c are constants not equal to zero is a quadratic trinomial. A good mnemonic for the factorization of a quadratic trinomial is the **PAFF** method.

PAFF method The letters P, A, F and F represent the four stages of the process:

P→**P**roduct;

A→**A**ddition;

F→**F**actors;

F→**F**actorize.

■ **Example 1.39** Factorize completely x^2+5x+6.

The stages of **PAFF** are as follows.

P→**P**roduct

Multiply the coefficient of x^2 by the constant term. It will represent a product.

$P=1\times6$

$=6$.

A→**A**ddition

Find the coefficient of x. It will represent an addition.

A$=5$.

F→**F**actors

Find two numbers that multiply to give P and add to give A.

P$=2\times 3$,

A$=2+3$.

F→Factorize

Use the factors obtained above to factorize the expression. **A** tells how the coefficient of x is going to be split.

$$\begin{aligned} x^2+5x+6 &= x^2+(2x+3x)+6 \\ &= (x^2+2x)+(3x+6) \\ &= x(x+2)+3(x+2) \\ &= (x+3)(x+2). \end{aligned}$$

■

Exercise 1.23 Factorize $6x^2+13x+5$ completely. ■

Solution **P**→Product

P$=6\times 5=30$.

A→Addition

A$=13$.

F→Factors

P$=3\times 10$,

A$=3+10$.

F→Factorize

$$\begin{aligned} 6x^2+13x+5 &= 6x^2+(3x+10x)+5 \\ &= (6x^2+3x)+(10x+5) \\ &= 3x(2x+1)+5(2x+1) \\ &= (3x+5)(2x+1). \end{aligned}$$

■

Exercise 1.24 Factorize the following expressions:

(1) $2x^2-3x+1$;

(2) x^2+x-6;

(3) $3q^2+7pq-6p^2$;

(4) $k^2-(2a-3b)k-6ab$. ■

Solution

(1) $$\begin{aligned} 2x^2-3x+1 &= 2x^2-2x-x+1 \\ &= (2x^2-2x)-(x-1) \\ &= 2x(x-1)-1(x-1) \end{aligned}$$

$$=(x-1)(2x-1).$$

(2) $x^2+x-6=x^2+3x-2x-6$
$=(x^2+3x)-(2x+6)$
$=x(x+3)-2(x+3)$
$=(x+3)(x-2).$

(3) $3q^2+7pq-6p^2=3q^2-2pq+9pq-6p^2$
$=(3q^2-2pq)+(9pq-6p^2)$
$=q(3q-2p)+3p(3q-2p)$
$=(q+3p)(3q-2p).$

(4) $k^2-(2a-3b)k-6ab=k^2-2ak+3bk-6ab$
$=k(k-2a)+3b(k-2a)$
$=(k-2a)(k+3b).$

■

Difference of two squares

An expression of the form a^2-b^2 is called a **difference of two squares.** It is useful to recognize this difference when factorizing expressions. For example, $16y^2-49$ is the difference between two perfect squares, written as $(4y)^2-(7)^2$. If a^2 and b^2 are two squares, then their difference can be written as $a^2-b^2=(a+b)(a-b)$.

Exercise 1.25 Factorize the following expressions:

(1) $4x^2-y^2$;

(2) $81a^2-4b^2$;

(3) $36p^2-49q^2$;

(4) $5-20y^2$;

(5) x^3-xy^2.

■

Solution

(1) $4x^2-y^2=(2x)^2-(y)^2$
$=(2x+y)(2x-y).$

(2) $81a^2-4b^2=(9a)^2-(2b)^2$
$=(9a+2b)(9a-2b).$

(3) $36p^2-49q^2=(6p)^2-(7q)^2$
$=(6p+7q)(6p-7q).$

(4) $5-20y^2=5(1-4y^2)$
$=5[1^2-(2y)^2]$
$=5(1+2y)(1-2y).$

(5) $x^3-xy^2=x(x^2-y^2)$
$=x(x+y)(x-y).$

■

1.5 Solving quadratic equations

Any equation of the form $ax^2+bx+c=0$, where a, b and c are constants and $a\neq 0$ is called a **quadratic equation**. For example, $2x^2+3x-1$ and $x^2+3=0$ are quadratic equations. A quadratic equation has two solutions or roots, not necessarily real, which may be equal or not. The main techniques for solving quadratic equations are given by:

(1) Factorization.

(2) Completion of the square.

(3) Graphical methods.

(4) Use of the quadratic formula.

Factorization

■ **Example 1.40** Solve the equation $(x+2)(x-3)=0$.

$$(x+2)(x-3)=0,$$
$$(x+2)=0 \text{ or } (x-3)=0,$$
$$x=-2 \text{ or } x=3.$$

■

Exercise 1.26 Solve the equation $x(x-5)=0$. ■

Solution

$$x(x-5)=0,$$
$$x=0 \text{ or } (x-5)=0,$$
$$x=0 \text{ or } x=5.$$

■

Exercise 1.27 Solve the equation $x^2-2x-3=0$. ■

Solution

$$x^2-2x-3=0,$$
$$x^2-3x+x-3=0,$$
$$x(x-3)+1(x-3)=0,$$
$$(x+1)(x-3)=0,$$
$$x=-1 \text{ or } x=3.$$

■

Ⓡ Sometimes, the quadratic equation is not given in the form $ax^2+bx+c=0$. When it is given in a different form, it should be put into the standard form first by suitably rearranging its terms. For example,

$x^2-5x=14$ becomes $x^2-5x-14=0$;

$3x^2-1=2x$ becomes $3x^2-2x-1=0$;

$x(x-1)=2$ becomes $x^2-x-2=0$;

$2-x^2=5x$ becomes $x^2+5x-2=0$.

Exercise 1.28 Solve the following equations:

(1) $x^2+2x=35$;

(2) $2x^2-x=3$;

(3) $x^2-5x=0$.

■

Solution

(1)
$$\begin{aligned} x^2+2x&=35,\\ x^2+2x-35&=0,\\ x^2+7x-5x-35&=0,\\ (x+7)(x-5)&=0,\\ x=-7 \text{ or } x&=5. \end{aligned}$$

(2)
$$\begin{aligned} 2x^2-x&=3,\\ 2x^2-x-3&=0,\\ 2x^2-3x+2x-3&=0,\\ x(2x-3)+1(2x-3)&=0,\\ (x+1)(2x-3)&=0,\\ x=-1 \text{ or } x&=\frac{3}{2}. \end{aligned}$$

(3)
$$\begin{aligned} x^2-5x&=0,\\ x(x-5)&=0,\\ x=0 \text{ or } x&=5. \end{aligned}$$

■

® Certain equations are not quadratic, but reduce to quadratic equations after a suitable variable change.

Exercise 1.29 Solve the following equations:

(1) $2^{2x}-9(2^x)+20=0$.

(2) $2^{2x+2}-9(2^x)+2=0$.

(3) $3^{2x}-30(3^x)+81=0$.

(4) $\left(\frac{1}{4}\right)^{3x^2+1}=128^x$.

■

Solution

(1) $2^{2x}-9(2^x)+20=0$, $\quad (2^{2x}=(2^x)^2)$

$$(2^x)^2-9(2^x)+20=0, \qquad \text{(substituting } 2^x=y)$$
$$y^2-9y+20=0, \qquad \text{(factorizing)}$$
$$(y-5)(y-4)=0,$$
$$y=4 \text{ or } y=5. \qquad \text{(remember that } y=2^x)$$
$$2^x=4=2^2\Rightarrow x=2,$$
$$2^x=5,$$
$$x=\log_2 5. \qquad \text{(taking } \log_2 \text{ of both sides)}$$

(2) $$2^{2x+2}-9(2^x)+2=0, \qquad (2^{2x+2}=2^2\times(2^x)^2)$$
$$2^2\times(2^x)^2-9(2^x)+2=0,$$
$$4\times(2^x)^2-9(2^x)+2=0, \qquad \text{(substituting } 2^x=p)$$
$$4p^2-9p+2=0, \qquad \text{(factorizing)}$$
$$(4p-1)(p-2)=0,$$
$$p=\frac{1}{4} \text{ or } p=2. \qquad \text{(remember that } p=2^x)$$
$$2^x=\frac{1}{4}=2^{-2}\Rightarrow x=-2,$$
$$2^x=2=2^1\Rightarrow x=1,$$
$$x=-2 \text{ or } x=1.$$

(3) $$3^{2x}-30(3^x)+81=0, \qquad (3^{2x}=(3^x)^2)$$
$$(3^x)^2-30(3^x)+81=0, \qquad \text{(substituting } 3^x=y)$$
$$y^2-30y+81=0, \qquad \text{(factorizing)}$$
$$(y-3)(y-27)=0,$$
$$y=3 \text{ or } y=27. \qquad \text{(remember that } y=3^x)$$
$$3^x=3=3^1\Rightarrow x=1,$$
$$3^x=27=3^3\Rightarrow x=3,$$
$$x=1 \text{ or } x=3.$$

(4) $$\left(\frac{1}{4}\right)^{3x^2+1}=128^x, \qquad \left(\frac{1}{4}=2^{-2}\right)$$
$$2^{-2(3x^2+1)}=2^{7x}, \qquad \text{(since } 2^7=128)$$
$$-2(3x^2+1)=7x, \qquad \text{(equating powers)}$$
$$6x^2+7x+2=0, \qquad \text{(factorizing)}$$
$$(2x+1)(3x+2)=0,$$
$$x=-\frac{1}{2} \text{ or } x=-\frac{2}{3}.$$

■

Completion of squares

■ **Example 1. 41** Using the completion of squares method, find the roots of the equation $x^2-6x+7=0$. The coefficient of x^2 is 1, therefore we add half of the coefficient of x, squared

$$x^2-6x=-7,$$

$$x^2-6x+(-3)^2=-7+(-3)^2,$$
$$(x-3)^2=-7+9,$$
$$(x-3)^2=2,$$
$$x-3=\pm\sqrt{2},$$
$$x=3\pm\sqrt{2},$$
$$x=3+\sqrt{2} \text{ or } x=3-\sqrt{2}.$$

■

Exercise 1.30 Find the roots of the equation $2x^2-3x-3=0$. ■

Solution First, make sure the coefficient of x^2 is 1 before completing the square. Now it is 2, and hence dividing by 2 we have $x^2-\frac{3}{2}x=\frac{3}{2}$. We may now add half of the coefficient of x, squared

$$x^2-\frac{3}{2}x+\left(-\frac{3}{4}\right)^2=\frac{3}{2}+\left(-\frac{3}{4}\right)^2,$$
$$\left(x-\frac{3}{4}\right)^2=\frac{3}{2}+\left(-\frac{3}{4}\right)^2,$$
$$\left(x-\frac{3}{4}\right)^2=\frac{33}{16},$$
$$x-\frac{3}{4}=\pm\sqrt{\frac{33}{16}}=\pm\frac{1}{4}\sqrt{33},$$
$$x=\frac{3}{4}\pm\frac{1}{4}\sqrt{33},$$
$$x=\frac{1}{4}(3+\sqrt{33}) \text{ or } x=\frac{1}{4}(3-\sqrt{33}).$$

■

Graphical methods

Exercise 1.31 Using the graphical method find the roots of the quadratic equation $x^2-8x+12=0$. ■

Solution When $y=0$ (on the x-axis), $x^2-8x+12=0$, and $a=1$, $b=-8$ and $c=12$. The axis of symmetry is

$$x=-\frac{b}{2a}=-\frac{-8}{2}=4.$$

Plugging $x=4$ into the function y gives

$$y=(4)^2-8(4)+12=-4,$$

which implies that the vertex is $(4,-4)$. For the x-intercepts, we have

$$x^2-8x+12=0\Rightarrow x=(2,0) \text{ and } (6,0),$$

and the y-intercept is $(0,12)$.

From Figure 1.7, we observe that the roots of the equation are $x=2$ and $x=6$.

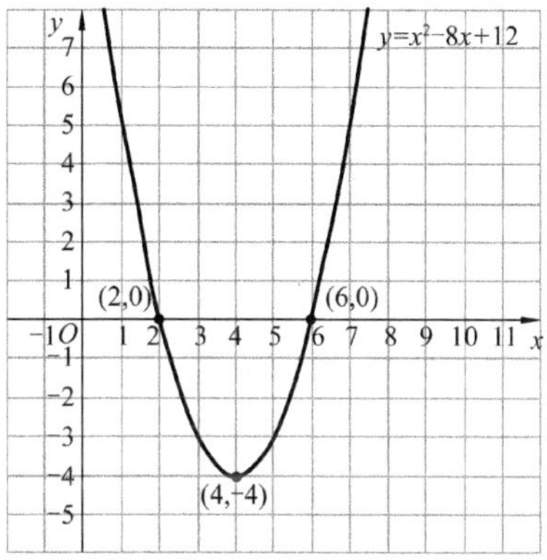

Figure 1.7 The parabola $y=x^2-8x+12$.

Exercise 1.32 Using the graphical method find the roots of the quadratic equation $-x^2-2x+8=0$.

Solution When $y=0$ (on the x-axis), $-x^2-2x+8=0$, and $a=-1$, $b=-2$ and $c=8$. The axis of symmetry is

$$x=-\frac{b}{2a}=-\frac{2}{2}=-1.$$

Plugging $x=-1$ into the function y gives

$$y=(-1)^2-2(-1)+8=9,$$

which implies that the vertex is $(-1,9)$. For the x-intercepts, we have

$$-x^2-2x+8=0\Rightarrow x=(2,0) \text{ and } (-4,0),$$

and the y-intercept $(0, 8)$.

From Figure 1.8, we observe that the roots of the equation are $x=2$ and $x=-4$.

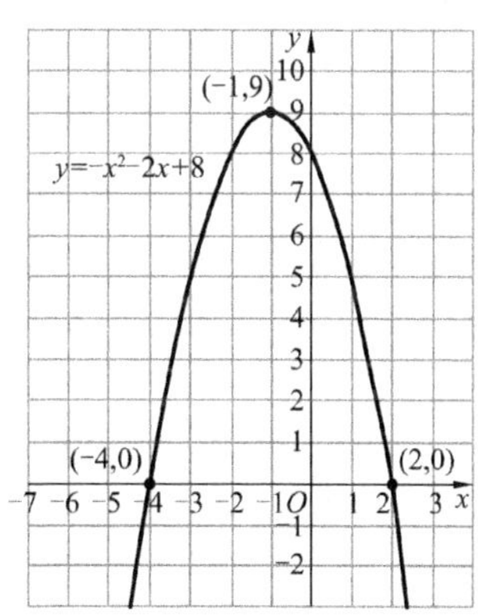

Figure 1.8 The parabola $y=-x^2-2x+8$.

Use of the quadratic formula

(i) Forming the equation when the roots are given

Given the roots of any quadratic equation, we can write the equation as

$$x^2-(\text{sum of roots})x+(\text{product of roots})=0.$$

Exercise 1.33 Find a quadratic equation with roots 3 and -4. ■

Solution Sum of roots: $3+(-4)=-1$. Product of roots: $3(-4)=-12$.

Therefore the equation is

$$x^2-(-1)x+(-12)=0,$$
$$x^2+x-12=0.$$ ■

Exercise 1.34 Find a quadratic equation with roots $-\frac{2}{3}$ and $-\frac{1}{4}$. ■

Solution Sum of roots: $-\frac{2}{3}+-\frac{1}{4}=-\frac{11}{12}$. Product of roots: $-\frac{2}{3}\times\left(-\frac{1}{4}\right)=\frac{1}{6}$.

Therefore the equation is

$$x^2-\left(-\frac{11}{12}\right)x+\left(\frac{1}{6}\right)=0,$$
$$12x^2+11x+2=0.$$ ■

(ii) Nature of roots of quadratic equations

The formula for finding the roots of the quadratic equation $ax^2+bx+c=0(a\neq 0)$, is

$$x=\frac{-b\pm\sqrt{b^2-4ac}}{2a},$$

in which $\Delta=b^2-4ac$ is called the **discriminant.** Knowing the sign of the discriminant allows deducing some properties of the roots without computing them.

1. If $b^2-4ac\geqslant 0$, then the equation has two real roots.
2. If $b^2-4ac>0$, then the equation has two real and different (or distinct) roots.
3. If $b^2-4ac=0$, then the equation has two real and equal roots.
4. If $b^2-4ac<0$, then the equation has no real root (i.e., the equation has complex roots).

■ **Example 1.42** Describe the nature of roots of the following equations:

(1) $2x^2-x-1=0$;

(2) $x^2+5x-1=0$;

(3) $3x^2-2x+4=0$;

(4) $2x^2-x+3=0$.

(1) Since $a=2, b=-1$ and $c=-1$,

$$b^2-4ac=(-1)^2-4(2)(-1)=9>0.$$

Therefore the equation has two real and different roots.

(2) Since $a=1$, $b=5$ and $c=-1$,

$$b^2-4ac=(5)^2-4(1)(-1)=29>0.$$

Therefore the equation has two real and different roots.

(3) Since $a=3$, $b=-2$ and $c=4$,

$$b^2-4ac=(-2)^2-4(3)(4)=-44<0.$$

Therefore the equation has complex roots.

(4) Since $a=2$, $b=-1$ and $c=3$,

$$b^2-4ac=(-1)^2-4(2)(3)=-23<0.$$

Therefore the equation has complex roots.

■

1.6 Simultaneous equations

Definition 1.6.1—Simultaneous equations. In mathematics, a set of **simultaneous equations**, also known as a **system of equations**, is a finite set of equations for which common solutions are sought.

1.6.1 Simultaneous linear equations

Linear equations may be put in the form $y=a_1x+a_0$, where a_1 and a_0 are real numbers and $a_1\neq 0$. Consequently, they are of degree 1. When graphed, these equations produce straight lines, which is why they are called linear.

A system of two linear equations can be solved using one of the following methods:

- substitution;
- elimination;
- graphical methods.

Substitution

Exercise 1.35 Solve the simultaneous equations $-2x+3y=19$ and $2x+y=1$. ■

Solution

$$\begin{cases} -2x+3y=19, & (1) \\ 2x+y=1. & (2) \end{cases}$$

$$\begin{aligned} y&=1-2x, && \text{(finding } y \text{ from (1))} \\ -2x+3(1-2x)&=19, && \text{(substituting } y=1-2x \text{ in (2))} \\ -2x+3-6x&=19, \\ -8x&=16, \\ x&=-2. \end{aligned}$$

To find y, substitute back the solution $x=-2$ into (2),

$\Rightarrow 2(-2)+y=1$

$\Rightarrow -4+y=1$

$\Rightarrow y=5.$

The solution of the simultaneous equations is then $x=-2$ and $y=5$.

■

Elimination

Exercise 1.36 Solve the simultaneous equations $2x-y=4$ and $x+y=5$. ■

Solution

$$\begin{cases} 2x-y=4, & (1) \\ x+y=5. & (2) \end{cases}$$

$3x+0=9,$ (adding equations (1) and (2) to eliminate y)

$x=3,$

$3+y=5$ (substituting $x=3$ into (2))

$y=2.$

The solution of the simultaneous equations is then $x=3$ and $y=2$.

■

Graphical methods

■ **Example 1.43** Solve the simultaneous equations $x+2y=7$ and $2x+3y=10$.

Using a **graphical method**, draw on the same axes the graphs of

$$x+2y=7 \quad \text{and} \quad 2x+3y=10.$$

As shown in Figure 1.9, the two lines cross. The point where they cross is on both lines, and so its coordinates satisfy both equations. We thereby need to find the coordinates of point of intersection. In this example, the point of intersection is $(-1, 4)$, and consequently the solution of the simultaneous equations is $x=-1$ and $y=4$.

■

Ⓡ The graphical method is also a useful tool for checking the algebraic methods.

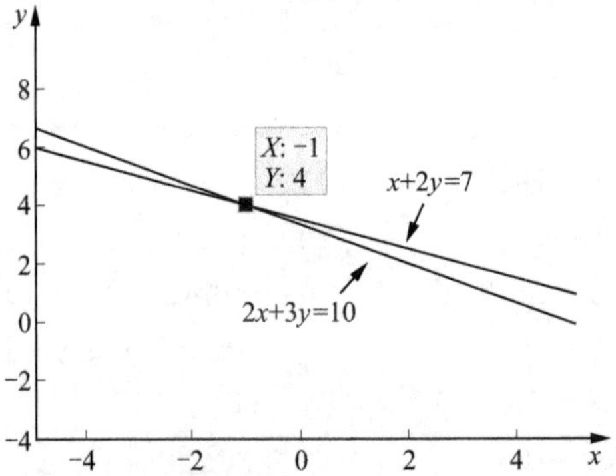

Figure 1.9 The graphs of $x+2y=7$ and $2x+3y=10$.

1.6.2 Simultaneous equations: linear and non-linear

A non-linear equation is (in the context of this book) a polynomial equation of degree 2 or higher. These equations can be solved using the same techniques as for linear equations. However, there is usually more than one solution.

Exercise 1.37 Solve the simultaneous equations $x-y+3=0$ and $x^2+y^2=29$. ■

Solution Using the substitution technique, find y from the first equation.

$$x-y+3=0 \Rightarrow y=x+3.$$

Substitute this result into the second equation,

$$2(x^2+3x-10)=0.$$

Factorize the quadratic polynomial as

$$2(x+5)(x-2)=0.$$

Therefore,

$$x=-5 \text{ or } x=2.$$

Substitute the new-found values of x back into the first equation and obtain the corresponding values of y.

$$y=-2 \text{ or } 5.$$

So there are two distinct solutions $x=-5$, $y=-2$ and $x=2$, $y=5$. Also, these solutions can be checked graphically as shown in Figure 1. 10. These can be written in coordinate form as $(-5,-2)$ and $(2, 5)$.

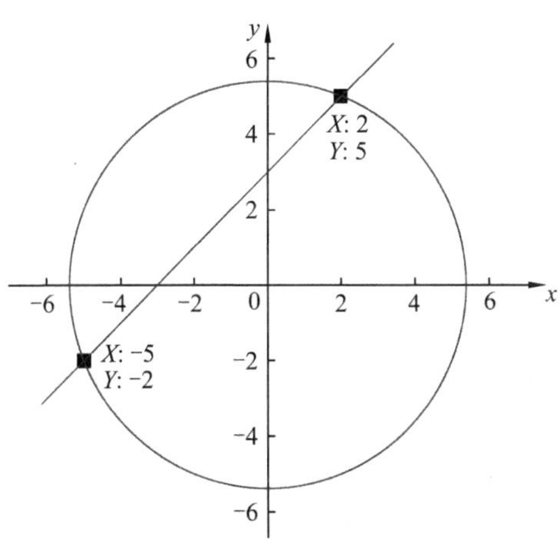

Figure 1. 10 The graphs of $x-y+3=0$ (straight line) and $x^2+y^2=29$ (circle).

■

1.7 Summary

Sets

Sets are conventionally denoted or represented by capital letters, for example by A, B, C. The components or objects of a set are called members or elements of the set and are enclosed in curly brackets ($\{\cdots\}$), being usually denoted by small letters. For example, if $A=\{2,3,4,5\}$, then 2 is a member or an element of A. The empty set (the set consisting of no elements at all) is denoted by $\varnothing$.

Set-theoretic operations

Various set-theoretic operations can be defined. The following are just a few examples.

(1) $A\cap B$ -**intersection** -objects that belong to both set A and set B.

(2) $A\cup B$ -**union** -objects that belong to set A or set B.

(3) $A\subset B$ -**proper subset** -all elements of A are also in B and A may not be equal to B.

(4) $A\subseteq B$ -**subset** -all elements of A are also in B and A may be equal to B.

(5) $A\not\subseteq B$ -**not subset** -the set A is not a subset of the set B.

(6) $A\supset B$ -**proper superset** -all elements of B are also in A and B may not be equal to A.

(7) $A\supseteq B$ -**supset** -all elements of B are also in A and B may be equal to A.

(8) $A\not\supseteq B$ -**not superset** -the set A is not a superset of the set B.

(9) $A=B$ -**equal** -both sets have the same members.

(10) $A\backslash B$ -**relative complement** -objects that belong to A and not to B.

(11) A^C-**complement** -all the objects that do not belong to set A.

Sets of numbers

A number is a mathematical object used to count, measure and label. The original examples are the natural numbers 1, 2, 3, 4 and so forth. A notational symbol that represents a number is called a numeral.

Natural numbers (N) : The counting numbers $\{0,1,2,3,\cdots\}$ are commonly called natural numbers.

Integers (Z) : Positive and negative counting numbers, as well as zero: $\{\cdots,-3,-2,-1,0,1,2,\cdots\}$. All natural numbers are integers, but not all integers are natural numbers.

Rational numbers (Q) : Numbers that can be expressed as a ratio of an integer to a non-zero integer. All integers are rational, but the converse is not true.

Real numbers (R) : Numbers that have decimal representations that have a finite or infinite sequence of digits to the right of the decimal point. They can be positive, negative, or zero. All rational numbers are real, but the converse is not true.

Irrational numbers (H) : Real numbers that are not rational.

Indices

Indices are a useful way of expressing large numbers in a simpler way. There are useful properties for manipulating expressions involving indices, called the laws of indices. The first two laws apply to the multiplication and division of powers with the same base.

The third law allows to raise a power of a number to another power. There are other properties who apply to the multiplication and division of powers with the same exponent, but we cannot use the laws of indices or other properties to manipulate expressions with both a different base and a different exponent.

The product law

$$a^m \times a^n = a^{m+n}.$$

The quotient law

$$\frac{a^m}{a^n} = a^{m-n}.$$

The power law

$$(a^m)^n = a^{mn}.$$

Power zero

$$a^0 = 1.$$

Power of product of two numbers

$$(ab)^n = a^n b^n.$$

Negative index

$$a^{-n} = \frac{1}{a^n},\ a \neq 0.$$

Power of quotient of two numbers

$$\left(\frac{a}{b}\right)^n = \frac{a^n}{b^n},\ b \neq 0.$$

Negative power of quotient of two numbers

$$\left(\frac{a}{b}\right)^{-n} = \left(\frac{b}{a}\right)^n = \frac{b^n}{a^n},\ a \neq 0,\ b \neq 0.$$

Fractional index

$$a^{\frac{1}{n}} = \sqrt[n]{a}.$$

Useful formulae

$$(a+b)(a-b) = a^2 - b^2.$$
$$(a+b)^2 = a^2 + 2ab + b^2.$$
$$(a-b)^2 = a^2 - 2ab + b^2.$$
$$(a+b)^3 = a^3 + 3a^2b + 3ab^2 + b^3.$$
$$(a-b)^3 = a^3 - 3a^2b + 3ab^2 - b^3.$$
$$(a+b+c)^2 = a^2 + b^2 + c^2 + 2ab + 2ac + 2bc.$$

Factorization

(1) For all polynomials, first factor out the greatest common factor (GCF).

(2) For a binomial, check to see if any of the following situations occur:

① difference of squares: $x^2 - y^2 = (x+y)(x-y)$;

② difference of cubes: $x^3 - y^3 = (x-y)(x^2+xy+y^2)$;

③ sum of cubes: $x^3+y^3=(x+y)(x^2-xy+y^2)$.

(3) For a trinomial, check to see whether it is either of the following forms:

① x^2+bx+c. If so, find two integers whose product is c and whose sum is b. For example, $x^2+8x+12=(x+2)(x+6)$ since $(2\times6=12)$ and $(2+6=8)$.

② ax^2+bx+c. if so, find two binomials so that the product of the first terms is ax^2 and the product of last terms is c and the sum of the outer and inner products is bx.

(4) For polynomials with four or more terms, regroup, factor each group, and then find a pattern as in Steps (1) through (3).

Solving quadratic equations

A quadratic equation is an equation that can be written in the form $ax^2+bx+c=0(a\neq 0)$. There are four distinct methods to solve quadratic equations:

(1) Factorizing.

(2) Completing the square.

(3) Graphical methods.

(4) Using the quadratic formula.

We shall obtain one of the following three possible solution sets:

(1) Two real distinct (i.e. not equal) solutions;

(2) A double root;

(3) Two complex solutions.

By checking the sign of the quantity b^2-4ac, which is called the discriminant.

(1) If $b^2-4ac>0$ then there are two real solutions of the quadratic equation.

(2) If $b^2-4ac=0$ then there are a double root of the quadratic equation.

(3) If $b^2-4ac<0$ then there are two complex solutions of the quadratic equation.

Simultaneous equations

A system including linear and non-linear equations can sometimes be solved simultaneously, using:

(1) substitution;

(2) elimination;

(3) graphical methods.

1.8 Further problems

Problem 1.1 If $P=\{\text{factors of } 42\}$ and $Q=\{\text{factors of } 54\}$, find $P\cap Q$.

Problem 1.2 If $D(n)$ denotes the set of factors of n, find

(1) $D(48)\cap D(64)$; (2) $D(12)\cap D(16)$.

Problem 1.3 Let U be the universal set of all prime numbers and let P and Q be subsets of U given by $P=\{x\mid x\in U, 3(x+1)\geqslant 2(x+10)\}$ and $Q=\{x\mid x\in U, 7<x<31\}$. List the elements of $P\cap Q$.

Problem 1.4 Let P and Q be subsets of the universal set $U=\{n \mid 1 \leqslant n \leqslant 10\}$, n being an integer. P consists of even numbers less than 7 and Q consists of prime numbers less than 10. List the members of

(1) $P \cup Q$; (2) $P \cap Q^C$;

(3) $P^C \cup Q^C$; (4) $P^C \cap Q^C$.

Problem 1.5 Let a, b, c, d be distinct. How many subsets does the set $A=\{a, b, c, d\}$ have?

Problem 1.6 The set A has 31 nonvoid subsets. How many elements does A have?

Problem 1.7 Sort out the following numbers into rational and irrational. Give reasons for your answers.

(1) $2.645\,376\,72\cdots$; (2) $\sqrt{1\,936}$;

(3) $\sqrt{27}$; (4) $2\sqrt{2}$;

(5) $-\frac{1}{7}$.

Problem 1.8 Evaluate the following:

(1) $2 \times \left(3\frac{1}{3}+1\frac{1}{6}\right)$; (2) $\left(3\frac{1}{2}+7\right) \div \left(4\frac{1}{3}-3\right)$;

(3) $2\frac{1}{4}+3\frac{4}{5}$; (4) $10 \div \left(3\frac{1}{2}+1\frac{1}{5}\right)$;

(5) $\frac{2}{5} \div 4$.

Problem 1.9 Arrange the following rational numbers in ascending order:

(1) 0.2, $\frac{1}{4}$, 30%, 10; (2) 0.15, $\frac{4}{6}$, 2, 40%.

Problem 1.10 Round 463.159 8 to the nearest

(1) whole number; (2) tenth;

(3) hundredth; (4) thousandth.

Problem 1.11 Express the following numbers in standard form

(1) 1 000 000; (2) 78 987 898;

(3) 100 065 656; (4) 0.000 034 241 4;

(5) 0.000 111 000 000 0; (6) 0.000 000 078 656 709 2.

Problem 1.12 Find all natural numbers n such that

(1) $\frac{1}{2}<\frac{n}{24}<\frac{5}{8}$; (2) $\frac{2}{5}>\frac{4}{3n-2}>\frac{1}{5}$.

Problem 1.13 Simplify the following, leaving your answer in standard form

(1) $\frac{0.016 \times 0.084}{0.48}$; (2) $\sqrt{\frac{0.81 \times 10^{-5}}{2.25 \times 10^{7}}}$;

(3) $\sqrt{\frac{0.004\,8 \times 0.81 \times 10^{-7}}{0.027 \times 0.04 \times 10^{6}}}$.

Problem 1.14 Let $A=6 \cdot 5^{n+1}+2 \cdot 5^{n+2}-17 \cdot 5^{n}$, $n \in \mathbf{N}$. Prove that $\frac{A}{3}$ is a natural number.

Problem 1.15 Find the value of x in the following equations:

(1) $3^{x-3}=81$; (2) $5^{2x+2}=25^2$.

(3) $3^x+3^{x+1}+3^{x+2}=351$.

Problem 1.16 If $a,b\in\mathbf{Q}^*$, $\frac{a}{b}=0.3$, determine:

(1) $\frac{b}{a}$; (2) $\frac{a}{3b}$;

(3) $\frac{a+b}{a}$; (4) $\frac{a^2+b^2}{ab}$.

Problem 1.17 After a price increase of 20%, a mobile phone costs 720 RMB. How much did it cost before the increase?

Problem 1.18 After two consecutive price decreases, the first of 5% and the second of 10%, a mobile phone costs 684 RMB. How much did it cost before the decreases?

Problem 1.19 Let $a=1+2+3+\cdots+49$. Evaluate a, prove that a is a perfect square and then determine $\sqrt{a}$.

Problem 1.20 Find x such that

(1) $\frac{0.25\times\sqrt{16}}{2^4-2^3}=\frac{x}{3^3-3^2-2}$; (2) $\frac{415-5\sqrt{2}}{x}=\frac{\sqrt{64+4\times3^2}}{1\frac{1}{5}\sqrt{\frac{25}{9}}}$.

Problem 1.21 Let $a,b,c,d\in\mathbf{Q}_+$ such that $\frac{1}{a+1}+\frac{1}{b+1}+\frac{1}{c+1}+\frac{1}{d+1}=2$. How much is $\frac{a}{a+1}+\frac{b}{b+1}+\frac{c}{c+1}+\frac{d}{d+1}$?

Problem 1.22 Factorize the following expressions:

(1) $3(x-1)+x(x-1)$; (2) $2x(x+5)-5(x+5)$;

(3) $8x(2x-1)-6(2x-1)^2$; (4) $(3x-5)^2-2(5-3x)+(x-5)(3x-5)$.

Problem 1.23 Let $a,b,c\in\mathbf{N}$.

(1) If $b+c=3$ and $ab+ac=12$, find a;

(2) If $b-c=5$ and $ab-ac=10$, find a;

(3) If $a=6$ and $ab-ac=18$, find $b-c$;

(4) If $a=4$ and $2a+3b+3c=23$, find $a^2+ab+ac$.

Problem 1.24 If $a-3b=3$, find the value of the expression

$$E=(2a-6b)+(3a-9b)^2+(4a-12b)^3.$$

Problem 1.25 If $a-4b-3=0$, find the value of the expression

$$E=(2a-8b-5)^{99}.$$

Problem 1.26 Find the value of the expressions:

(1) $(2a-3b+4c)^{15}+(3b-2a-4c)^{15}$;

(2) $(3a-4b-5c)^{81}+(-3a+4b+5c)^{81}$.

Problem 1.27 Using the formula $(a+b)(a-b)=a^2-b^2$, compute:

(1) $(4+2\sqrt{3})(4-2\sqrt{3})$; (2) $(5\sqrt{3}+2\sqrt{5})(5\sqrt{3}-2\sqrt{5})$;

(3) $\left(\frac{1}{2}x+\frac{3}{4}\right)\left(\frac{1}{2}x-\frac{3}{4}\right)$.

Problem 1.28 Using the formula $(a+b)(a-b)=a^2-b^2$, compute:

(1) $(x-1)(x+1)(x^2+1)$; (2) $(x-2)(x+2)(x^2+4)$.

(3) $(\sqrt{2}x-\sqrt{3}y)(\sqrt{2}x+\sqrt{3}y)(2x^2+3y^2)$.

Problem 1.29 Let $a\in\mathbf{N}$. Prove that the following expressions are perfect squares:

(1) $a(a+4)+4$; (2) $a(a-6)+9$;

(3) $(a-2)(a-4)+1$; (4) $(a^2+a)(a^2+a+4)+4$.

Problem 1.30 Solve the following quadratic equations by completing the square:

(1) $x^2-6x-16=0$; (2) $-3x^2+8x+7=0$.

Problem 1.31 Find a quadratic equations with roots $\frac{1}{3}$ and $\frac{3}{4}$.

Problem 1.32 Solve the following simultaneous equations:

(1) $3x+y=7$ and $x-2y=5$;

(2) $\frac{3}{2}x-\frac{1}{2}y=5$ and $\frac{5}{2}x+\frac{2}{3}y=12$.

Problem 1.33 In amoney box of $5c$ and $10c$ coins, there are 71 coins. Their total value is 5.60 RMB. Find how many coins are of each type.

Problem 1.34 Solve the simultaneous equations

(1) $x+2y=1$ and $x^2-6y^2=1$; (2) $x+y=2$ and $x^2+y^2=2$.

Problem 1.35 Find two numbers whose sum equals 5 and whose product equals -14.

—— Answers ——

Problem 1.1 $P\cap Q=\{1,2,3,6\}$.

Problem 1.2 (1) $D(48)\cap D(64)=\{1,2,4,8,16\}$. (2) $D(12)\cap D(16)=\{1,2,4\}$.

Problem 1.3 $P\cap Q=\{17,19,23,29\}$.

Problem 1.4 (1) $P\cup Q=\{2,3,4,5,6,7\}$.

(2) $P\cap Q^C=\{4,6\}$.

(3) $P^C\cup Q^C=\{1,3,4,5,6,7,8,9,10\}$.

(4) $P^C\cap Q^C=\{1,8,9,10\}$.

Problem 1.5 $2^4=16$ subsets.

Problem 1.6 5 elements.

Problem 1.7 (1) 2.645 376 72… (irrational)

(2) $\sqrt{1\,936}$ (irrational)

(3) $\sqrt{27}$ (irrational)

(4) $2\sqrt{2}$ (irrational)

(5) $-\frac{1}{7}$ (rational)

Problem 1.8 (1) $2\times\left(3\frac{1}{3}+1\frac{1}{6}\right)=2\times\left(\frac{10}{3}+\frac{7}{6}\right)=2\times\left(\frac{27}{6}\right)=\frac{27}{3}=9$.

(2) $\left(3\frac{1}{2}+7\right)\div\left(4\frac{1}{3}-3\right)=10\frac{1}{2}\div1\frac{1}{3}=\frac{21}{2}\div\frac{4}{3}=\frac{63}{8}$.

(3) $2\frac{1}{4}+3\frac{4}{5}=\frac{9}{4}+\frac{19}{5}=\frac{45+76}{20}=\frac{121}{20}$.

(4) $10\div\left(3\frac{1}{2}+1\frac{1}{5}\right)=10\div\left(\frac{7}{2}+\frac{6}{5}\right)=10\div\left(\frac{35+12}{10}\right)=10\div\frac{47}{10}=\frac{100}{47}$.

(5) $\frac{2}{5}\div4=\frac{2}{5}\times\frac{1}{4}=\frac{1}{10}$.

Problem 1.9 (1) $0.2<\frac{1}{4}<30\%<10$. (2) $0.15<40\%<\frac{4}{6}<2$.

Problem 1.10 (1) 463 to the nearest whole number.

(2) 463.2 to the nearest tenth.

(3) 463.16 to the nearest hundreath.

(4) 463.160 to the nearest thousandth.

Problem 1.11 (1) $1\,000\,000=1\times10^6$.

(2) $78\,987\,898=7.898\,789\,8\times10^7$.

(3) $100\,065\,656=1.000\,656\,56\times10^8$.

(4) $0.000\,034\,241\,4=3.424\,14\times10^{-5}$.

(5) $0.000\,111\,000\,000\,0=1.110\,000\,000\times10^{-4}$.

(6) $0.000\,000\,078\,656\,709\,2=7.865\,670\,92\times10^{-8}$.

Problem 1.12 (1) $n=13,14$. (2) $n=5,6,7$.

Problem 1.13 (1) $\frac{0.016\times 0.084}{0.48}=2.8\times 10^{-3}$.

(2) $\sqrt{\frac{0.81\times 10^{-5}}{2.25\times 10^{7}}}=6\times 10^{-7}$.

(3) $\sqrt{\frac{0.0048\times 0.81\times 10^{-7}}{0.027\times 0.04\times 10^{6}}}=6\times 10^{-7}$.

Problem 1.14 Since $A=30\cdot 5^n+50\cdot 5^n-17\cdot 5^n=63\cdot 5^n$, it is obvious that $\frac{A}{3}=21\cdot 5^n$, which implies A that $\frac{A}{3}$ is natural number.

Problem 1.15 (1) 7. (2) 1. (3) 3.

Problem 1.16 (1) $\frac{10}{3}$. (2) 0.1. (3) $\frac{13}{3}$. (4) $\frac{103}{30}$.

Problem 1.17 600 RMB.

Problem 1.18 800 RMB.

Problem 1.19 $2a=(1+49)+(2+48)+\cdots+(48+2)+(49+1)=2\,450$, so $a=1\,225$. Also, one notes that $(35)^2=1\,225$.

Problem 1.20 (1) 2. (2) 41.

Problem 1.21 2.

Problem 1.22 (1) $(3+x)(x-1)$.

(2) $(2x-5)(x+5)$.

(3) $2(2x-1)(-2x+3)$.

(4) $4(3x-5)(x-2)$.

Problem 1.23 (1) $a=4$. (2) $a=2$. (3) $b-c=3$. (4) $a^2+ab+ac=36$.

Problem 1.24 $E=1\,815$.

Problem 1.25 $E=1$.

Problem 1.26 (1) 0. (2) 0.

Problem 1.27 (1) 4. (2) 55. (3) $\frac{1}{4}x^2-\frac{9}{16}$.

Problem 1.28 (1) x^4-1. (2) x^4-16. (3) $4x^4-9y^4$.

Problem 1.29 (1) $(a+2)^2$. (2) $(a-3)^2$. (3) $(a-3)^2$. (4) $(a^2+a+2)^2$.

Problem 1.30 (1) $x=8, -2$. (2) $x=\frac{4\pm\sqrt{37}}{3}$.

Problem 1.31 $x^2-\frac{13}{12}x+\frac{1}{4}=0$.

Problem 1.32 (1) $x=\frac{19}{7}$, $y=-\frac{8}{7}$. (2) $x=\frac{112}{27}$, $y=\frac{22}{9}$.

Problem 1.33 There are 30 5c coins and 41 10c coins.

Problem 1.34 (1) $x=1, y=0$ and $x=5, y=-2$. (2) $x=1, y=1$.

Problem 1.35 7 and -2.

Chapter 2 Coordinate Geometry

2.1 Introduction

Any point P in the x-y plane can be described by an ordered pair of numbers (x, y), where x is the x-**coordinate** (or **abscissa**) and the y is the y-**coordinate** (or **ordinate**). For instance, the point $A(3,4)$ has 3 as the x-coordinate and 4 as the y-coordinate as shown in Figure 2.1 below. The **origin** O has coordinates $(0,0)$.

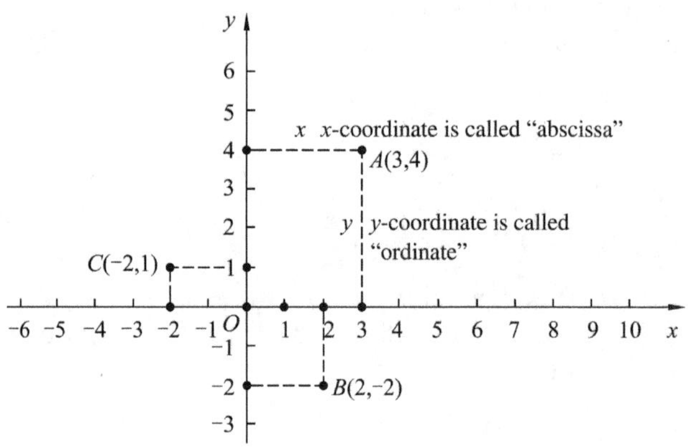

Figure 2.1 Axis of geometry.

® All points of the x-axis (and only those points) have ordinate 0. All points on the y-axis (and only those points) have abscissa 0.

2.1.1 The distance between two points

If $P(a,b)$ and $Q(c,d)$ are two points, then the **distance** between P and Q (or, in other words, the **length of the line segment** PQ) is given by:

$$|PQ| = \sqrt{(c-a)^2 + (d-b)^2}.$$

® One of the squares contains the difference of the x-coordinates (the horizontal change), while the other contains the difference of y-coordinates (the vertical change). It doesn't matter which point you start with, P or Q.

Exercise 2.1 Find the distance between the following points:
(1) $A(1, 3)$ and $B(4,7)$;
(2) $P(-1,-3)$ and $Q(4, 9)$;
(3) $X(5,-6)$ and $Y(-2,5)$;
(4) $S(4,6)$ and $R(12,6)$.
■

Solution

(1) $|AB|=\sqrt{(4-1)^2+(7-3)^2}=\sqrt{3^2+4^2}=\sqrt{25}=5$;

(2) $|PQ|=\sqrt{[4-(-1)]^2+[9-(-3)]^2}=\sqrt{5^2+12^2}=\sqrt{169}=13$;

(3) $|XY|=\sqrt{[-2-5]^2+[5-(-6)]^2}=\sqrt{(-7)^2+(-11)^2}=\sqrt{170}$;

(4) $|SR|=\sqrt{(12-4)^2+(6-6)^2}=\sqrt{8^2+0^2}=\sqrt{64}=8$.

■

2.1.2 The midpoint of a line segment

Suppose that M is the midpoint of the line segment joining the points $A(x_1, y_1)$ and $B(x_2, y_2)$, represented in Figure 2.2 below.

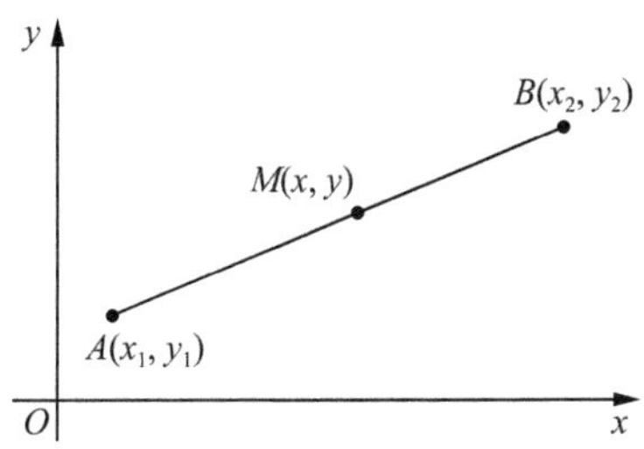

Figure 2.2 Midpoint of a line segment.

Then the coordinates of M are given by:

$$x=\frac{1}{2}(x_1+x_2),\ y=\frac{1}{2}(y_1+y_2).$$

Ⓡ The midpoint M is, in some sense, the mean between A and B. Consequently, its coordinates are the arithmetic mean of the respective coordinates of A and B.

Exercise 2.2 Find the coordinates of the midpoint of the line segment determined by the following points:
(1) $C(1,3)$ and $D(5,7)$;
(2) $A(-3,1)$ and $B(7,3)$;
(3) $A(5, 7)$ and $M(0,2)$;
(4) $X(-1,-4)$ and $Y(2,6)$.
■

Solution

(1) $\left(\frac{1}{2}(1+5), \frac{1}{2}(3+7)\right)=(3,5)$;

(2) $\left(\frac{1}{2}(-3+7), \frac{1}{2}(1+3)\right)=(2,2)$;

(3) $\left(\frac{1}{2}(5+0), \frac{1}{2}(7+2)\right)=\left(\frac{5}{2},\frac{9}{2}\right)$;

(4) $\left(\frac{1}{2}(-1+2), \frac{1}{2}(-4+6)\right)=\left(\frac{1}{2},1\right)$.

■

2.1.3 The slope of a line joining two points

As shown in Figure 2.3, if $A(x_1, y_1)$ and $B(x_2, y_2)$ are two points such that AB is not vertical, then the **slope** or **gradient** of the line joining A and B is given by:

$$m=\frac{\text{Difference in } y\text{-values}}{\text{Difference in } x\text{-values}}=\frac{\text{Vertical change}}{\text{Horizontal change}}=\frac{y_2-y_1}{x_2-x_1}=\tan\theta,$$

where θ is the **angle of incline.**

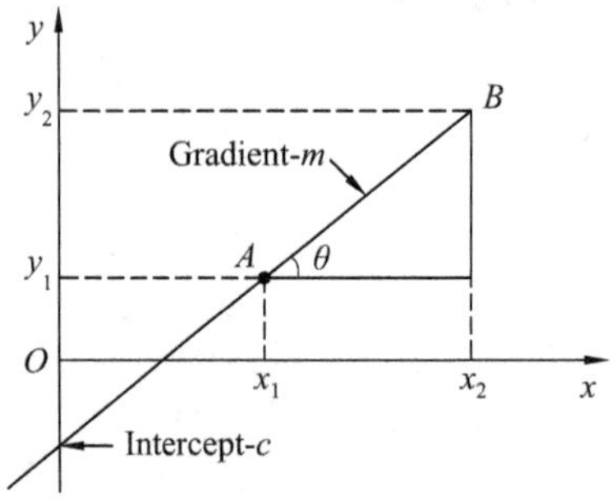

Figure 2.3 Gradient of a straight line.

Ⓡ The slope of a line does not change along that line. To compute the slope of a line determined by 2 points, it does not matter which points you take or start with. The slope of a vertical line is undefined, since there is no change in the x-values along this line and the denominator of the fraction defining the slope is 0.

Ⓡ Colloquially, the slope of a (nonvertical) line can be defined as $\frac{\text{rise}}{\text{run}}$. A line with positive slope goes up. A line with negative slope goes down. A line with slope 0 is horizontal.

Exercise 2.3 Find the slope of the line determined by the following two points:
(1) $P(1, 1)$ and $Q(7, 2)$;
(2) $A(2,1)$ and $B(5,5)$;
(3) $M(3,-2)$ and $N(-3,4)$;
(4) $X(3,-2)$ and $Y(-5,-8)$.

■

Solution

(1) $\frac{2-1}{7-1}=\frac{1-2}{1-7}=\frac{1}{6}$.

(2) $\frac{5-1}{5-2}=\frac{1-5}{2-5}=\frac{4}{3}$.

(3) $\frac{4-(-2)}{-3-3}=\frac{6}{-6}=-1$.

(4) $\frac{-8-(-2)}{-5-3}=\frac{-6}{-8}=\frac{3}{4}$.

■

Exercise 2.4 Find the slopes of the lines with the following equations:

(1) $2x+4y=3$;

(2) $2x-8y+3=0$.

■

Solution

(1) $2x+4y=3$

$4y=-2x+3$ (making y the subject)

$y=-\frac{1}{2}x+\frac{3}{4}$ (the coefficient of x is $-\frac{1}{2}$)

Therefore, the slope of the line is $-\frac{1}{2}$.

(2) $2x-8y+3=0$

$8y=2x+3$ (making y the subject)

$y=\frac{1}{4}x+\frac{3}{8}$ (the coefficient of x is $\frac{1}{4}$)

Therefore, the slope of the line is $\frac{1}{4}$.

■

2.2 Equations of a straight line

The equations of straight lines can take various forms, depending on what is known about those lines.

Horizontal lines

On a horizontal line the y-coordinate of the points stays constant. Consequently, the equation of a horizontal line is

$$y=\text{constant}.$$

■ **Example 2.1** Find the equation of the horizontal line passing through $A(1,2)$.

All points on this horizontal line have the same y-coordinate, which is the y-coordinate of A. The equation of the line is then $y=2$.

■

Vertical lines

On a vertical line, the x-coordinate of the points stays constant. Consequently, the equation of a vertical line is

$$x=\text{constant}.$$

■ **Example 2.2** Find the equation of the vertical line passing through $A(3, 5)$.

All points on this vertical line have the same x-coordinate, which is the x-coordinate of A. The equation of the line is then $x=3$.

■

Point-slope equation

The equation of a straight line through the point $M(a,b)$ with a given slope m is

$$y-b=m(x-a) \text{ or } y=m(x-a)+b.$$

® The x-coefficient in the equations above is the slope of the line.

■ **Example 2.3** Find the equation of a line with slope 2 which passes through $A(1,3)$.

The equation is

$$y-3=2(x-1)\Rightarrow y-3=2x-2\Rightarrow y=2x+1.$$

■

Exercise 2.5 Find the equation of the line which passes through $A(1,2)$ and $B(2, 1)$.

■

Solution The slope of the line is $\dfrac{1-2}{2-1}=-1$. Since the line passes through A, using the point-slope equation gives

$$y-2=(-1)(x-1)\Rightarrow y-2=-x+1\Rightarrow y=-x+3.$$

It is easy to see that using B instead of A leads us to

$$y-1=(-1)(x-2)=y-1=-x+2=-y=-x+3,$$

that is, we obtain the same equation.

■

Slope-intercept equation

The equation of a line with a given slope m and a given y-intercept b is

$$y=mx+b.$$

® The slope-intercept equation is obtained from the point-slope equation by setting $a=0$. The point-slope equation can be written taking any point on a line, the same equation being ultimately obtained no matter which point one starts with. The slope-intercept equation is unique from the start, since the slope and y-intercept are both uniquely defined.

Intercept-intercept equation

Assume a straight line intersects x-axis at $(a, 0)$ and y-axis at $(0, b)$. Then its equation is

$$\frac{x}{a}+\frac{y}{b}=1 \quad (a\neq 0,\ b\neq 0),$$

which also can be written as

$$xb+ya=ab.$$

Ⓡ The latter form is somewhat more general as it allows either a or b to be 0. a and b are defined as x-intercept and y-intercept of the linear function. These are signed distances from the points of intersection of the line with the axes.

■ **Example 2.4** Find the equation of the line which intersects the x-axis at $A(-2,0)$ and the y-axis at $B(0,3)$. Determine its slope and intercept.

The intercept-intercept equation is

$$\frac{x}{-2}+\frac{y}{3}=1\Rightarrow -\frac{3}{2}x+y=3\rightarrow y=\frac{3}{2}x+3,$$

and consequently the slope is $\frac{3}{2}$ and the y-intercept is 3.

■

Exercise 2.6 Find the equations of the following straight lines:

(1) a straight line that passes through the point $M(0,1)$ and has slope 3;

(2) a straight line that passes through the points $A(2,\ 3)$ and $B(3,6)$;

(3) a straight line that passes through the points $C(0,-1)$ and $D(-2,0)$.

■

Solution

(1) The equations is

$$y-1=3(x-0),$$

that is,

$$3x-y+1=0.$$

(2) Since the line passes through $A(2,3)$ and $B(3,6)$, its slope is

$$\frac{6-3}{3-2}=3.$$

The equation of the straight line is then

$$y-3=3(x-2),$$

that is,

$$3x-y-3=0.$$

(3) Since the line passes through the points $C(0,-1)$ and $D(-2,0)$, its intercept-intercept equation is

$$\frac{x}{-2}+\frac{y}{-1}=1,$$

that is,

$$x+2y+2=0.$$

■

General equation

Inspecting all line equations written above, one sees that the general equation of a straight line has a linear form

$$Ax+By+C=0,$$

where A, B and C are constants, and A, B are not both 0.

The general form is not the most useful to use when writing the equation of a line (note that it has three parameters instead of two). However, this form does unify all other equations of a straight line.

■ **Example 2.5** Find out which of the points $A(1, 3)$ and $B(2, 1)$ belong to the line of equation $2x-3y-1=0$. Since $2\times1-3\times3-1=-8\neq0$, the coordinates of A do not verify the given equation and consequently A does not belong to the line. Since $2\times2-3\times1-1=0$, the coordinates of B verify the given equation and consequently B belongs to the line. ■

■ **Example 2.6** Find the slope and the y-intercept of the line with equation $4x-2y+3=0$. To find the slope and the intercept, one should put the equation in the slope-intercept form. It is seen that

$$4x-2y+3=0\Rightarrow2y=4x+3\Rightarrow y=2x+\frac{3}{2},$$

and consequently the slope is 2 and the y-intercept is $\frac{3}{2}$. ■

Parallel and perpendicular lines

Definition 2.2.1—Parallel lines. In geometry, **parallel lines** are lines in a plane that do not intersect each other, that is, two lines in a plane that do not meet are said to be parallel.

Notation 2.1 If the lines m and n are parallel, we denote that by $m/\!/n$.

Geometric conditions for parallelism

Given two straight lines m and n in the Euclidean space, the following properties are equivalent.

(1) Line m is in the same plane as line n but it does not intersect n (parallelism).

(2) Every point on line m is located at exactly the same (minimum) distance from line n (equidistance).

(3) Lines m and n are in the same plane and, when both are intersected by a third straight line (a transversal), the corresponding angles of intersection with the transversal are congruent (see Figure 2.4).

(4) Lines m and n are in the same plane and, when both are intersected by a transversal, the sum of the two interior angles equals π.

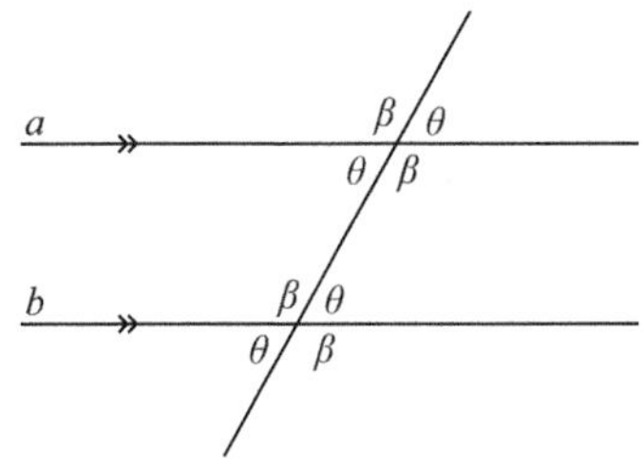

Figure 2. 4 Lines a and b are parallel if and only if the transversal determines congruent corresponding angles θ, shown here both to the right of the transversal, one above and adjacent to line a and the other above and adjacent to line b.

Analytic conditions for parallelism

Two non-vertical lines $y=m_1x+b_1$ and $y=m_2x+b_2$ are parallel if and only if they have equal angles of incline, and consequently equal slopes, but different intercepts, that is

$$m_1=m_2 \text{ and } b_1\neq b_2.$$

Note also that the lines coincide if and only if they have both equal slopes and intercepts, that is,

$$m_1=m_2 \text{ and } b_1=b_2.$$

■ **Example 2. 7** Find the equation of a line which passes through $A(3,1)$ and is parallel to the line of equation $y=3x+4$.

The line of equation $y=3x+4$ has slope 3. Since the line to be determined is parallel to this line, it also has slope 3, its equation being

$$y-1=3(x-3)\Rightarrow y-1=3x-9\Rightarrow y=3x-8.$$

■

When the equations of the lines are given in the general form $a_1x+b_1y+c_1=0$ and $a_2x+b_2y+c_2=0$, $a_2, b_2, c_2\neq 0$ and a_1, b_1 not both 0, the lines are parallel if and only if

$$\frac{a_1}{a_2}=\frac{b_1}{b_2}\neq\frac{c_1}{c_2}.$$

Note that, in this case, the equation of the second line can be put in the form

$$a_1x+b_1y+\frac{c_2a_1}{a_2}=0$$

after a division with $\frac{a_2}{a_1}$, leading to similar coefficients for x and y as those of the first equation.

■ **Example 2. 8** Find a such that the lines with equations $3x+ay+4=0$ and $x+y+2=0$ are parallel. The lines are parallel if and only if

$$\frac{3}{1}=\frac{a}{1}\neq\frac{4}{2}\Rightarrow a=3.$$

■

The distance between two parallel lines

Given the equations of two non-vertical parallel lines, $y=mx+b_1$ and $y=mx+b_2$, $b_1\neq b_2$, the distance between them is

$$d=\frac{|b_2-b_1|}{\sqrt{1+m^2}}.$$

■ **Example 2.9** Find the distance between the lines with equations $y=3x+1$ and $y=3x-2$. The lines are parallel since both have slope 3, while having different intercepts, 1 and -2, respectively. The distance between them is then

$$d=\frac{|-2-1|}{\sqrt{1+3^2}}=\frac{3}{\sqrt{10}}=\frac{3\sqrt{10}}{10}.$$

■

When the equations of the parallel lines are given in the general form: $ax+by+c_1=0$ and $ax+by+c_2=0$ (a, b not both 0, $c_1\neq c_2$), the distance between the lines is

$$d=\frac{|c_2-c_1|}{\sqrt{a^2+b^2}}.$$

■ **Example 2.10** Find the distance between the lines with equations $2x+y-3=0$ and $6x+3y+1=0$. The lines are parallel since is then

$$\frac{2}{6}=\frac{1}{3}\neq\frac{-3}{1}.$$

The second equation can be put in the form $2x+y+\frac{1}{3}=0$ via division by 3. The distance between the lines

$$d=\frac{\left|\frac{1}{3}-(-3)\right|}{\sqrt{2^2+1^2}}=\frac{\frac{10}{3}}{\sqrt{5}}=\frac{10}{3\sqrt{5}}=\frac{10\sqrt{5}}{15}=\frac{2\sqrt{5}}{3}.$$

■

Definition 2.2.2—Perpendicular lines. A line is said to be **perpendicular** to another line if the two lines intersect at a right angle.

Notation 2.2 If the lines m and n are perpendicular, we denote that by $m\perp n$.

Analytic conditions for perpendicularity

Two non-vertical lines $y=m_1x+b_1$ and $y=m_2x+b_2$ are perpendicular if and only if the slope of a line is equal to the negative reciprocal of the other slope, that is, if and only if

$$m_1m_2=-1.$$

■ **Example 2.11** Let $A(4,9)$, $B(4,-1)$, $C(-4,3)$ be three points. Find the equation of a line which passes through A and is perpendicular to BC.

The line determined by B and C has slope $\frac{3-(-1)}{-4-4}=-\frac{1}{2}$. Denoting by m the slope of the line to be determined, it folows that $m\times\left(-\frac{1}{2}\right)=-1$, and consequently $m=2$. The equation sought after is $y-9=2(x-4)$, that is, $y=2x+1$.

■

When the equations of the lines are given in the general form $a_1x+b_1y+c_1=0$ and $a_2x+b_2y+c_2=0$, the lines are perpendicular if and only if

$$a_1a_2+b_1b_2=0.$$

■ **Example 2.12** Find a such that the lines with equations $x+2y+1=0$ and $2x-ay+3=0$ are perpendicular.

The lines are perpendicular if and only if

$$1\times 2+2\times(-a)=0\Rightarrow 2-2a=0\Rightarrow a=1.$$

■

Definition 2.2.3—Foot of a perpendicular. Let A be a point and let m be a line. If B is the point of intersection of m and the unique line through A that is perpendicular to m, then B is called the **foot** of this perpendicular through A.

Perpendicularity easily extends to segments and rays. For example, as shown in Figure 2.5, a line segment AB is perpendicular to a line segment CD if, when each is extended in both directions to form an infinite line, these two resulting lines are perpendicular in the sense above. In symbols, $AB\perp CD$ means line segment AB is perpendicular to line segment CD.

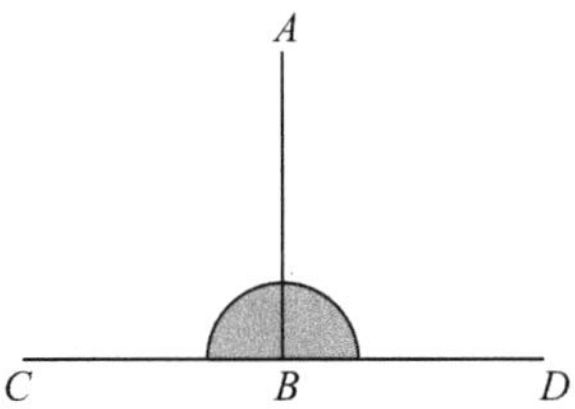

Figure 2.5 The line segment *AB* is perpendicular to the line segment *CD* because the two angles it determines are 90° each. The point *B* is the foot of the perpendicular from *A* to the segment *CD*.

® Since the diagram above can be in any orientation, the foot is not necessarily at the bottom.

Relationship to parallel lines

If two lines are both perpendicular to a third line, all of the angles formed along the third line are right angles. Therefore, in Euclidean geometry, any two lines that are both perpendicular to a third line are parallel to each other. Conversely, if one line is perpendicular to a second line, it is also perpendicular to any line parallel to that second line.

Exercise 2.7 Find the equations of the following lines:

(1) A line that passes through the point $A(1,3)$ and is perpendicular to the line $2y-3x+1=0$;

(2) A line that passes through the point $B(3,-2)$ and is parallel to the line $5x-y+3=0$.

■

Solution

(1) $2y-3x+1=0$,

$$m=\frac{3}{2},$$

$$y=\frac{3}{2}x-\frac{1}{2}. \qquad \text{(making } y \text{ the subject)}$$

$$\text{Gradient of the line }=-\frac{1}{m} \qquad \text{(since the lines are perpendicular)}$$

$$=-\frac{1}{3/2}$$

$$=-\frac{2}{3}.$$

The slope-intercept equation of the line is of the form

$$y-y_1=m(x-x_1),$$

with $(x_1, y_1)=(1,3)$ and $m=-\frac{2}{3}$. Hence,

$$y-3=-\frac{2}{3}(x-1),$$

that is,

$$2x+3y-11=0.$$

(2) $5x-y+3=0$,

$$m=5,$$

$$y=5x+3. \qquad \text{(making } y \text{ the subject)}$$

Gradient of the line$=5$. (since the two lines are parallel)

The slope-intercept equation of the line is of the form

$$y-y_1=m(x-x_1),$$

with $(x_1, y_1)=(3,-2)$ and $m=5$. Hence,

$$y+2=5(x-3),$$

that is,

$$5x-y-17=0.$$

■

2.3 Inequalities

2.3.1 Symbols

In mathematics, an inequality is a relation that holds between two values when they are different.

- The notation $a\neq b$ means that a is **not equal to** b. It does not say that one is greater than the other.
- The notation $a<b$ means that a is **less than** b.
- The notation $a>b$ means that a is **greater than** b.
- The notation $a\leqslant b$ means that a is **less than or equal to** b (or equivalently **not greater than** b or **at most** b). The notation $a\geqslant b$ means that a is **greater than or equal to** b (or, equivalent, **not less than** b or **at least** b).

® The notation $a<b$ ($a>b$) may also be read as "a is **strictly less than** b" ("a is **strictly greater than** b").

2.3.2 Properties

1. **Transitive property.** When we line up inequalities in order, we can "jump over" the middle inequality. For instance, if $a<b$ and $b<c$, then $a<c$.

■ **Example 2.13** If Alex is older than Billy and Billy is older than Carol, then Alex must be older than Carol also.

■

2. **Reversal property.** We can swap a and b over, if we make sure the symbol still "points at" the smaller value (see Figure 2.6). For instance, if $a<b$, then $b>a$; if $a>b$, then $b<a$.

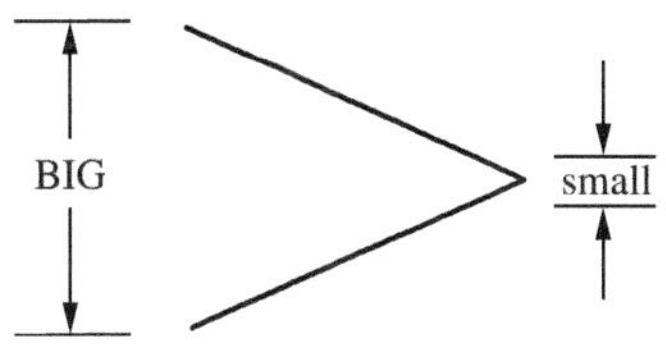

Figure 2.6 The symbol "points at" the smaller value.

■ **Example 2.14** Alex is older than Billy, so Billy is younger than Alex.

■

3. **Law of trichotomy.** The "Law of trichotomy" says that only one of the following relation is true:

$$a>b,\ a<b,\ a=b.$$

■ **Example 2.15** Either Alex has more money than Billy, or Alex has less money than Billy, or they both have the same amount of money.

■

4. **Addition and subtraction.** Adding c to both sides of an inequality just shifts everything along, and the inequality stays the same (see Figure 2.7).

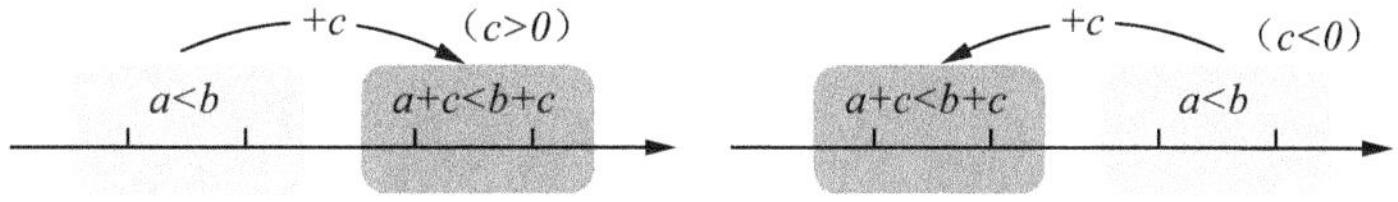

Figure 2.7 If $a<b$, then $a+c<b+c$.

■ **Example 2.16** Alex has less coins than Billy. If both Alex and Billy get 3 more coins each, Alex will still have less coins than Billy.

■

5. **Multiplication and division.** When we multiply both a and b by a **positive** number, the inequality stays the **same.** But when we multiply both a and b by a **negative** number, the inequality **swaps over** (see Figure 2.8).

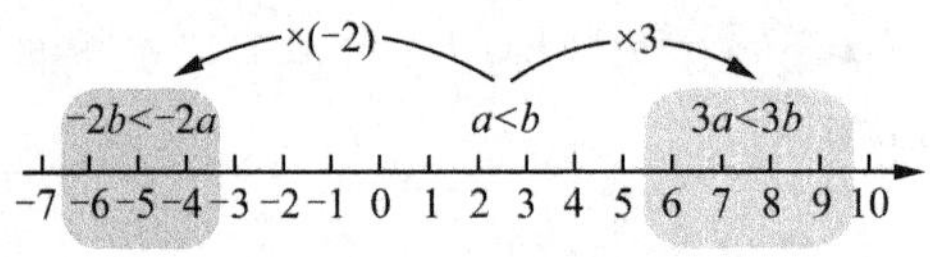

Figure 2.8 $a<b$ **becomes** $(-2)a>(-2)b$ **after multiplying by** (-2)**. But the inequality stays the same after multiplying by 3, becoming** $3a<3b$**.**

■ **Example 2.17** Alex's score of 3 is lower than Billy's score of 7. If both Alex and Billy manage to double their scores, Alex's new score of 6 will still be lower than Billy's new score of 14.

■

6. **Additive inverse.** As we just saw, putting **minuses** in front of a and b **changes the direction** of the inequality. For instance, if $a<b$, then $-a>-b$ and if $a>b$, then $-a<-b$.

■ **Example 2.18** Alex has more money than Billy, and so Alex is ahead. But a new law says "all your money is now a debt you must repay with hard work". So now Alex is worse off than Billy.

■

7. **Multiplicative Inverse.** Taking the **reciprocal** (1/value) of both a and b can **change the direction** of the inequality. For instance, when a and b are **both** positive or both negative, if $a<b$, then $\frac{1}{a}>\frac{1}{b}$ and if $a>b$, then $\frac{1}{a}<\frac{1}{b}$. But when **either** a **or** b is negative (not both) the direction stays the **same.**

■ **Example 2.19** Alex and Billy both complete a journey of 12 kilometers. Alex runs at 6 km/h and Billy walks at 4 km/h. Alex's speed is greater than Billy's speed. But Alex's time is less than Billy's time: 2 hours<3 hours.

■

8. **Non-negative property of squares.** A square of a number is greater than or equal to zero. For instance: $(-3)^2=9>0$, $0^2=0$.

9. **Square root property.** Taking a square root will not change the inequality (but only when both a and b are greater than or equal to zero). For instance, if $4<9$, then $\sqrt{4}=2<\sqrt{9}=3$.

2.3.3 Solving linear inequalities involving one variable

Solving inequalities is very much like solving equations. We mostly perform the same steps, but we must also pay attention to the **direction of the inequality.**

For example, solve

$$x+3<0.$$

If one has given us the corresponding equation "$x+3=0$" instead, we would have known how to solve it: we would have subtracted 3 from both sides. We can do the same thing here. Then the solution is $x<-3$.

The formatting of the above answer is called "inequality notation", because the solution is written as an inequality. This is probably the simplest of the solution notations, but there are three others with which you might need to be familiar.

(1) "Set notation" writes the solution as a set of points. The above solution would be written in set notation as

$$\{x \mid x \text{ is a real number, } x<-3\},$$

which reads as "the set of all x-values, such that x is a real number, and x is less than minus three".

(2) "Interval notation" writes the solution as an interval (that is, as a section or length along the number line). The above solution, "$x<-3$", would be written as "$(-\infty, -3)$", which reads as "the interval from negative infinity to minus three".

(3) The last "notation" is more of an illustration. You may be directed to "graph" the solution (see Figure 2.9).

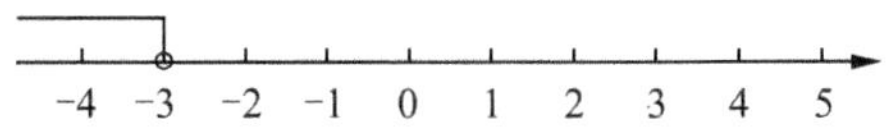

Figure 2.9 Since -3 is not included in the solution, we mark this point with an open dot pointing in the direction of the rest of the solution interval.

■ **Example 2.20** Solve the inequality $2x+1\geqslant 7$.

$$2x+1\geqslant 7,$$
$$2x\geqslant 6,$$
$$x\geqslant 3.$$

The solution is then pictured in Figure 2.10.

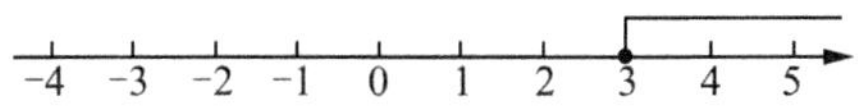

Figure 2.10 Since 3 is included in the solution, we mark this point with a closed dot pointing in the direction of the rest of the solution interval.

■

Exercise 2.8 Solve the inequalities:

(1) $5x+4<3x+10$.

(2) $\frac{6+x}{3}\geqslant x+4$.

■

Solution

(1) $5x+4<3x+10$,

$5x-3x<10-4$, (collecting similar terms)

$2x<6$, (dividing by 2)

$x<3$.

(2) $\frac{6+x}{3}\geqslant x+4$, (multiplying by 3)

$6+x\geqslant 3(x+4)$,

$$x-3x\geqslant 12-6, \qquad \text{(collecting similar terms)}$$
$$-2x\geqslant 6, \qquad \text{(dividing by}-2)$$
$$x\leqslant -3.$$

■

An inequality with "x^2" as the highest-degree term is called a **quadratic inequality**. Solving a quadratic inequality is slightly more complicated than solving a linear inequality.

Exercise 2.9 Solve $x^2-3x+2>0$. ■

Solution First, we have to find the x-intercepts of the associated quadratic, because the intercepts are where $y=x^2-3x+2$ is equal to zero. Graphically, the inequality like this is asking us to find where the graph is above or below the x-axis. It is simpler to find where it actually crosses the x-axis and consequently we shall start there. Factoring, we obtain

$$x^2-3x+2=(x-2)(x-1)=0,$$

so $x=1$ or $x=2$.

Then the graph crosses the x-axis at 1 and 2, and the number line is divided into the intervals $(-\infty,1)$, $(1,2)$, and $(2,+\infty)$. Between the x-intercepts, the graph is either above the axis (and thus the quadratic is positive), or below the axis (and thus the quadratic is negative). There are two different algebraic ways of checking for positivity or negativity on each interval, as seen below.

1. **Test-point method.** The intervals between the x-intercepts are $(-\infty,1)$, $(1,2)$, and $(2,+\infty)$, respectively. We shall pick a point (any point) inside each interval and determine the sign of the quadratic at that point.

2. **Factor method.** We first factorize the quadratic as $x^2-3x+2=(x-2)(x-1)$. Subsequently, we consider each of these factors separately. The factor $x-1$ is positive for $x>1$. Similarly, $x-2$ is positive for $x>2$. Thinking about the sign rule, to compute the sign on $y=x^2-3x+2$, we only really need to know the signs on the factors. Finally, the solution of $x^2-3x+2>0$ are the two intervals

$$(-\infty,1)\cup(2,+\infty).$$

■

Exercise 2.10 Solve $x^3+9x^2+23x+15\geqslant 0$. ■

Solution First, we factor to find the zeros

$$x^3+9x^2+23x+15=(x+5)(x+3)(x+1)=0.$$

To solve by the Test-point method, we would pick a sample point in each interval, the intervals being $(-\infty,-5)$, $(-5,-3)$, $(-3,-1)$ and $(-1,+\infty)$. As you can see, if the polynomial or rational function has many factors, the Test-point method can become quite time-consuming.

To solve by the Factor method, we would solve each factor for its positivity: $x+5>0$ for $x>-5$; $x+3>0$ for $x>-3$; $x+1>0$ for $x>-1$. Then one analyzes the sign of each factor and obtains that the solution is

$$[-5,-3]\cup[-1,+\infty).$$

■

Exercise 2.11 Solve $\frac{x}{x-3}\leqslant 2$. ■

Solution

$$\frac{x}{x-3}\leqslant 2,$$

$$\frac{x}{x-3}-2\leqslant 0.$$

Now we need to convert to a common denominator.

$$\frac{x}{x-3}-\frac{2(x-3)}{x-3}\leqslant 0,$$

$$\frac{-x+6}{x-3}\leqslant 0,$$

$$\frac{x-6}{x-3}\geqslant 0.$$

Note that x cannot actually be equal to 3, so this endpoint will not be included in any solution interval (even though this is an "or equal to" inequality), but we need the value in order to figure out what the intervals are. In this case, our intervals are $(-\infty,3)$, $(3,6]$, and $[6,+\infty)$. Note the use of brackets to indicate that 6 can be included in the solution, but that 3 cannot.

Finally, the solution is

$$(-\infty,\ 3)\cup[6,+\infty).$$

■

2.3.4 Solving linear inequalities involving two variables

Figure 2.11 shows a graph of a linear inequality $y\leqslant x+2$ (the shaded area is where y is less than or equal to $x+2$).

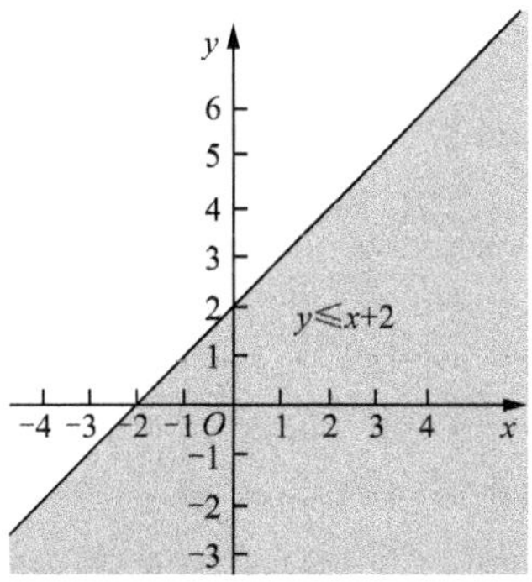

Figure 2.11 The inequality $y\leqslant x+2$.

How to graph a linear inequality

First, graph the "equals" line, then shade in the correct area. There are three steps:

(1) Rearrange the equation so "y" is on the left and everything else on the right.

(2) Plot the "$y=$" line (make it a solid line for $y\leqslant$ or $y\geqslant$, and a dashed line for $y<$ or $y>$).

(3) Shade above the line for a "greater than" or below the line for a "less than".

■ **Example 2.21** Solve $2y-x\leqslant 6$.

We shall need to rearrange this one so that y is on its own on the left

$$2y\leqslant x+6\Rightarrow y\leqslant \frac{x}{2}+3.$$

Now plot $y=\frac{x}{2}+3$ (as a solid line because $y\leqslant$ includes equal to) (see Figure 2.12).

Shade the area below (because y is less than or equal to) (see Figure 2.13).

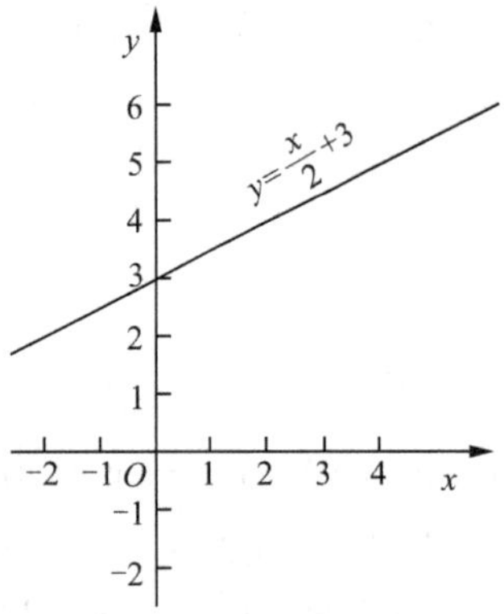

Figure 2.12 The equality $y=\frac{x}{2}+3$.

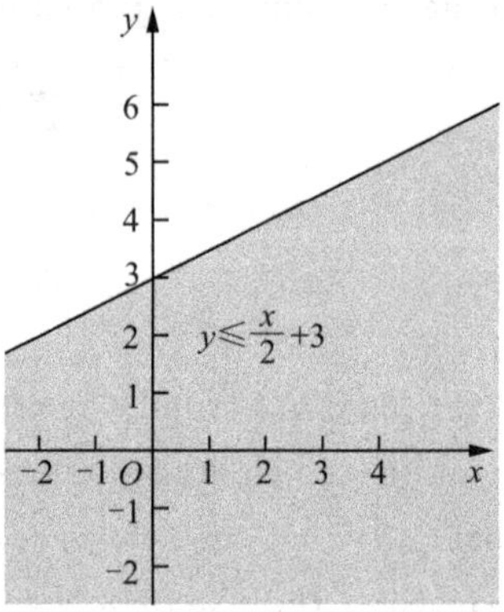

Figure 2.13 The inequality $y\leqslant\frac{x}{2}+3$.

■

2.4 The equation of a circle

Definition 2.4.1—Circle. A **circle** is a simple closed shape in Euclidean geometry. It is the set of all points in a plane that are at a given distance, called the **radius** from a given point, called the **centre**. Equivalently, a circle is the curve traced out by a point that moves so that its distance from a given point is constant.

Ⓡ Since the radius is a distance, it is always taken to be positive. Also, the plural of "radius" is "radii".

2.4.1 Terminology

There is a (comprehensive) terminology to be known when studying circles from a mathematical viewpoint (see Figure 2.14).

Annulus The ring-shaped region bounded by two concentric circles.

Arc Any connected part of the circle.

Chord A line segment whose endpoints lie on the circle.

Circumference The length of one circuit along the circle, or the distance around the circle.

Diameter A line segment whose endpoints lie on the circle and which passes through the centre; or the length of such a line segment, which is the largest distance between any two points on the circle. It is a special case of a chord, namely the longest chord, and it is twice the radius.

Disc The region of the plane bounded by a circle.

Lens The intersection of two discs.

Passant A coplanar straight line that does not touch the circle.

Sector A region bounded by two radii and an arc lying between the radii.

Segment A region, not containing the centre, bounded by a chord and an arc lying between the chord's endpoints.

Secant An extended chord, a coplanar straight line cutting the circle at two points.

Semicircle An arc that extends from one of a diameter's endpoints to the other. In non-technical common usage it may mean the diameter, arc, and its interior, a two dimensional region, that is technically called a half-disc. A half-disc is a special case of a segment, namely the largest one.

Tangent A coplanar straight line that touches the circle at a single point.

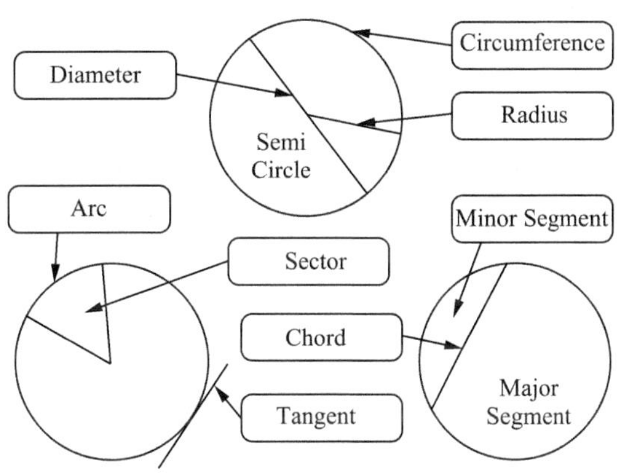

Figure 2. 14 Circle terminology.

2.4.2 Cartesian coordinates

The standard form

In an x-y Cartesian coordinate system, the circle with centre coordinates (a, b) and radius r is the set of all points (x, y) such that

$$(x-a)^2+(y-b)^2=r^2.$$

This equation, known as the **standard form** of the equation of a circle, follows from the Pythagorean theorem applied to any point on the circle: as shown in Figure 2. 15, the radius is the hypotenuse of a right-angled triangle whose other sides are of length $|x-a|$

and $|y-b|$.

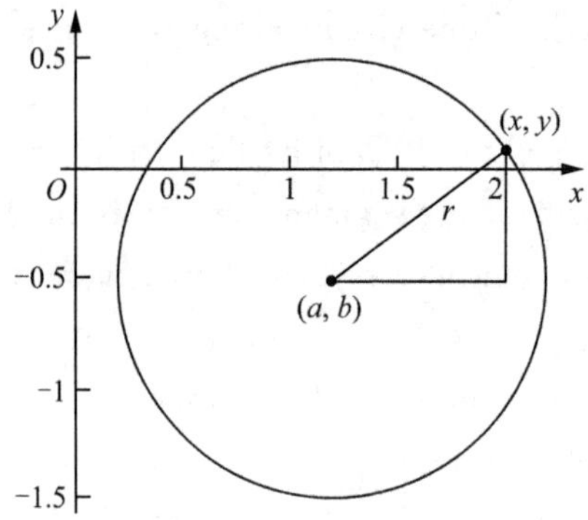

Figure 2.15 Circle of radius r and centre (a, b).

■ **Example 2.22** Find the equation of the circle with centre $C(2, -3)$ and radius $r=5$. Since $a=2$, $b=-3$ and $r=5$, the equation is $(x-2)^2+(y+3)^2=25$.

■

Exercise 2.12 Find the radius and centre of the following circle

$$(x-3)^2+(y+2)^2=36.$$

■

Solution Compare $(x-3)^2+(y+2)^2=36$ with $(x-a)^2+(y-b)^2=r^2$. The equation is in the same form with $a=3$, $b=-2$ and $c=6$. So $(x-3)^2+(y+2)^2=36$ is the equation of a circle with radius 6 and centre coordinates $(3,-2)$.

■

Exercise 2.13 (1) The point $(k,2)$ lies on the circle $x^2+y^2=20$. Find the value of k.

(2) The point $(k,0)$ lies on the circle with center coordinates $(5,2)$ and radius $\sqrt{8}$. Find the value of k.

■

Solution

(1) Substitute $(k,2)$ into the equation of the circle. Then

$$k^2+4=20 \Rightarrow k=\pm 4.$$

Consequently, there are two possible values of k.

(2) The equation for a circle of radius $\sqrt{8}$ and center coordinates $(5,2)$ is given by

$$(x-5)^2+(y-2)^2=(\sqrt{8})^2.$$

Now, substitute $x=k$ and $y=0$ into this equation and obtain

$$(k-5)^2+(0-2)^2=8,$$

which implies that

$$(k-5)^2=4 \Rightarrow k=5\pm 2.$$

So, $k=7$ or $k=3$.

■

Exercise 2.14 Find the equation of the circle for which the extremities of a diameter are $A(3,2)$ and $B(-1, 6)$.

■

Solution The center of the circle is the midpoint of the line segment AB, while the radius is half the length of the diameter AB. The coordinates of the midpoint are

$$\left(\frac{3+(-1)}{2}, \frac{2+6}{2}\right)=(1,4),$$

while the length of AB is

$$\sqrt{(-1-3)^2+(6-2)^2}=\sqrt{32}=4\sqrt{2}.$$

It follows that the radius is $r=2\sqrt{2}$ and since $a=1$ and $b=4$, the equation of the circle is

$$(x-1)^2+(y-4)^2=(2\sqrt{2})^2 \Rightarrow (x-1)^2+(y-4)^2=8.$$

■

The reduced form

If the circle is centred at the origin $O(0,0)$, then its equation is given by

$$x^2+y^2=r^2.$$

■ **Example 2.23** Find the equation of the circle centered at the origin which passes through $A(3,4)$. The radius of the circle is the length of the line segment OA, equal to

$$\sqrt{(3-0)^2+(4-0)^2}=\sqrt{3^2+4^2}=\sqrt{25}=5.$$

It follows that the equation of the circle is

$$x^2+y^2=5^2 \Rightarrow x^2+y^2=25.$$

■

The general form

Expanding the parentheses in the standard form, we observe that the equation of a circle has the general form

$$x^2+y^2+Ax+By+C=0,$$

A, B and C being all constants.

■ **Example 2.24** The equation of a circle in standard form is

$$(x+11)^2+(y-9)^2=16.$$

What is the general form of the equation of this circle?

We expand $(x+11)^2+(y-9)^2=16$ and obtain $x^2+22x+121+y^2-18y+81=16$. We then collect similar terms and end up with this general form

$$x^2+y^2+22x-18y+186=0.$$

■

■ **Example 2.25** The equation of a circle in the general form is

$$x^2+y^2-6x-4y+8=0.$$

Put this equation into the standard form and determine the centre and radius of this circle.

Rearranging terms, we obtain

$$x^2+y^2-6x-4y+8=0 \Rightarrow x^2-6x+y^2-4y+8=0.$$

Adding half of the coefficient of x, squared, and half the coefficent of y, also squared, one obtains

$$x^2-6x+3^2+y^2-4y+2^2+8=3^2+2^2$$
$$\Rightarrow (x-3)^2+(y-2)^2+8=13$$
$$\Rightarrow (x-3)^2+(y-2)^2=5.$$

The center is $C(3,2)$ and the radius is $r=\sqrt{5}$.

■

2.5 Summary

Cartesian coordinates

Any point P in the x-y plane can be described by an ordered pair of numbers (x,y), where x is the x-**coordinate** (or **abscissa**) and the y is the y-**coordinate** (or **ordinate**).

The distance between two points

If $P(a,b)$ and $Q(c,d)$ are two points, then the **distance** between P and Q (or, in other words, the **length of the line segment** PQ) is given by:

$$|PQ|=\sqrt{(c-a)^2+(d-b)^2}.$$

The midpoint of a line segment

Suppose that M is the midpoint of the line segment joining the points $A(x_1,y_1)$ and $B(x_2,y_2)$. Then the coordinates of M are given by:

$$x=\frac{1}{2}(x_1+x_2),\ y=\frac{1}{2}(y_1+y_2).$$

The slope of a line joining two points

If $A(x_1,y_1)$ and $B(x_2,y_2)$ are two points such that AB is not vertical, then the **slope** or **gradient** of the line joining A and B is given by:

$$m=\frac{\text{Difference in } y\text{-values}}{\text{Difference in } x\text{-values}}=\frac{\text{Vertical change}}{\text{Horizontal change}}=\frac{y_2-y_1}{x_2-x_1}=\tan\theta,$$

where θ is the **angle of incline.**

Equations of vertical or horizontal lines

On a horizontal line the y-coordinate of the points stays constant. Consequently, the equation of a horizontal line is

$$y=\text{constant}.$$

On a vertical line, the x-coordinate of the points stays constant. Consequently, the equation of a vertical line is

$$x=\text{constant}.$$

Point-slope equation

The equation of a straight line through the point $M(a,b)$ with a given slope m is

$$y-b=m(x-a) \text{ or } y=m(x-a)+b.$$

Slope-intercept equation

The equation of a line with a given slope m and a given y-intercept b is

$$y=mx+b.$$

General equation

Inspecting all line equations written above, one sees that the general equation of a straight line has a linear form as given below

$$Ax+By+C=0,$$

where A,B and C are constants, and A,B are not both 0.

Intercept-intercept equation

Assume a straight line intersects x-axis at $(a,0)$ and y-axis at $(0,b)$. Then its equation is

$$\frac{x}{a}+\frac{y}{b}=1 \quad (a\neq 0,\ b\neq 0),$$

which also can be written as

$$xb+ya=ab.$$

Parallel lines

In geometry, **parallel lines** are lines in a plane which do not meet each other. The "parallel" symbol is $/\!/$.

Geometric conditions for parallelism

Given two straight lines m and n in the Euclidean space, the following properties are equivalent.

(1) Line m is in the same plane as line n but it does not intersect n (parallelism).

(2) Every point on line m is located at exactly the same (minimum) distance from line n (equidistance).

(3) Lines m and n are in the same plane and, when both are intersected by a third straight line (a transversal), the corresponding angles of intersection with the transversal are congruent.

(4) Lines m and n are in the same plane and, when both are intersected by a

transversal, the sum of the two interior angles equals π.

Analytic conditions for parallelism

Two non-vertical lines $y=m_1x+b_1$ and $y=m_2x+b_2$ are parallel if and only if they have equal angles of incline, and consequently equal slopes, but different intercepts, that is

$$m_1=m_2 \text{ and } b_1\neq b_2.$$

The lines coincide if and only if they have both equal slopes and intercepts, that is,

$$m_1=m_2 \text{ and } b_1=b_2.$$

When the equations of the lines are given in the general form $a_1x+b_1y+c_1=0$ and $a_2x+b_2y+c_2=0$, $a_2, b_2, c_2\neq 0$ and a_1, b_1 not both 0, the lines are parallel if and only if

$$\frac{a_1}{a_2}=\frac{b_1}{b_2}\neq\frac{c_1}{c_2}.$$

The distance between two parallel lines

Given the equations of two non-vertical parallel lines, $y=mx+b_1$ and $y=mx+b_2$, $b_1\neq b_2$, the distance between them is

$$d=\frac{|b_2-b_1|}{\sqrt{1+m^2}}.$$

When the equations of the parallel lines are given in the general form: $ax+by+c_1=0$ and $ax+by+c_2=0$ (a, b not both 0, $c_1\neq c_2$), the distance between the lines is

$$d=\frac{|c_2-c_1|}{\sqrt{a^2+b^2}}.$$

Perpendicular lines

A line is said to be **perpendicular** to another line if the two lines intersect at a right angle. The "perpendicular" symbol is "$\perp$".

Analytic conditions for perpendicularity

Two non-vertical lines $y=m_1x+b_1$ and $y=m_2x+b_2$ are perpendicular if and only if the slope of a line is equal to the negative reciprocal of the other slope, that is, if and only if

$$m_1m_2=-1.$$

When the equations of the lines are given in the general form $a_1x+b_1y+c_1=0$ and $a_2x+b_2y+c_2=0$, the lines are perpendicular if and only if

$$a_1a_2+b_1b_2=0.$$

Foot of a perpendicular

Let A be a point and let m be a line. If B is the point of intersection of m and the unique line through A that is perpendicular to m, then B is called the **foot** of this perpendicular through A.

Inequalities symbols

- The notation $a \neq b$ means that a is **not equal to** b. It does not say that one is greater than the other.
- The notation $a < b$ means that a is **less than** b.
- The notation $a > b$ means that a is **greater than** b.
- The notation $a \leqslant b$ means that a is **less than or equal to** b.
- The notation $a \geqslant b$ means that a is **greater than or equal to** b.

How to graph a linear inequality involving two variables

First, graph the "equals" line, then shade in the correct area. There are three steps:

(1) Rearrange the equation so "y" is on the left and everything else on the right;

(2) Plot the "$y=$" line (make it a solid line for $y \leqslant$ or $y \geqslant$, and a dashed line for $y <$ or $y >$).

(3) Shade above the line for a "greater than" or below the line for a "less than".

Circle

A **circle** is the set of all points in a plane that are at a given distance, called the **radius** from a given point, called the **centre.**

The standard equation

In an x-y Cartesian coordinate system, the **standard form** of the equation of a circle with centre coordinates (a, b) and radius r is

$$(x-a)^2+(y-b)^2=r^2.$$

The reduced form

If the circle is centred at the origin $(0,0)$, then its equation simplifies to

$$x^2+y^2=r^2.$$

The general form

The equation of a circle has the general form

$$x^2+y^2+Ax+By+C=0,$$

A, B and C being constants.

2.6 Further problems

Problem 2.1 Find the midpoints of the sides of the $\triangle ABC$ with vertices $A(1,2)$, $B(3,6)$, $C(-1,-4)$.

Problem 2.2 Find the distance between the origin O and the point $A(-3,-4)$.

Problem 2.3 Find the perimeter of the $\triangle ABC$ with vertices $A(-2,-1)$, $B(-3,2)$, $C(-5,0)$.

Problem 2.4 Consider the points $A(3,2)$, $B(4,3)$, $C(5,0)$.

(1) Find the length of the sides AB, AC, BC.

(2) Prove that the triangle ABC is right-angled.

Problem 2.5 Does the line of equation $y=-x+1$ rise, or descend? What about the line of equation $y=3x-2$?

Problem 2.6 Prove that the lines of equation $2x-3y+4=0$ and $4x-6y=-8$, respectively, coincide.

Problem 2.7 Which are the slope and the y-intercept of the line with equation $y=-2x+4$? What about the x-intercept?

Problem 2.8 Find a,b such that the points $A(0,-2)$ and $B(2,0)$ belong to the line of equation $y=ax+b$.

Problem 2.9 Find a such that the point $A(2a-1,a^2)$ belongs to the line of equation $x-y+1=0$.

Problem 2.10 Find the x-intercept and the y-intercept of the line with equation $3x+4y-12=0$.

Problem 2.11 Prove that the points $A(1,1)$, $B(2,4)$, $C(3,7)$ are collinear (lie on the same line).

Problem 2.12 Find a such that the points $A(3,5)$, $B(5,3)$, $C(a, a)$ are collinear.

Problem 2.13 Find the equation of the line which passes through $O(0,0)$ and the midpoint of the line segment determined by the points $A(4,0)$ and $B(0,2)$.

Problem 2.14 Find the vertices of the $\triangle ABC$ given the coordinates of its midpoints $M(-1,2)$, $N(4,-1)$, $P(2,-2)$.

Problem 2.15 Find a such that the lines of equations $2x+3y+1=0$, $3x+y-2=0$ and $x+y+a=0$ are concurrent (intersect at the same point).

Problem 2.16 Prove that the lines of equations $3x+4y+1=0$ and $6x+8y+12=0$ are parallel and find the distance between them.

Problem 2.17 Solve the inequalities:

(1) $3x+1\geqslant 5$; (2) $2y+3\leqslant 7$.

Problem 2.18 Solve the inequalities:

(1) $3x+5>x+10$; (2) $\frac{6+x}{2}\leqslant x+1$.

Problem 2.19 Solve the inequalities:

(1) $(3x-9)(2x+2)(5x+10)<0$; (2) $(8-4x)(3x+6)(2x-2)\geqslant 0$.

Problem 2.20 Solve the inequalities:

(1) $\frac{4x-8}{3x+9}\leqslant 0$; (2) $\frac{3x+1}{8-2x}>0$.

Problem 2.21 Solve the quadratic inequalities:

(1) $x^2+2x<0$; (2) $-x^2+4x-3\geqslant 0$.

Problem 2.22 Solve the inequalities:

(1) $y-2x\geqslant 6$; (2) $2y+x\leqslant 6$.

Problem 2.23 (1) The point $(1,k)$ lies on the circle $x^2+y^2=10$. Find the value of k.

(2) The point $(0,k)$ lies on the circle with center $(2,2)$ and radius $\sqrt{8}$. Find the value of k.

Problem 2.24 The equation of a circle in standard form is

$$(x+4)^2+(y-1)^2=2^2.$$

What is the equation of the circle in general form?

Problem 2.25 The equation of a circle in general form is

$$x^2+y^2-4x-2y-20=0.$$

What is the equation of the circle in standard form?

Answers

Problem 2.1 $(2, 4)$, $(1,1)$, $(0,-1)$.

Problem 2.2 5.

Problem 2.3 $2\sqrt{2}(\sqrt{5}+1)$.

Problem 2.4 (1) $AB=\sqrt{2}$, $BC=\sqrt{10}$, $AC=2\sqrt{2}$. (2) It follows from $AB^2+AC^2=BC^2$ that the triangle ABC is right-angled.

Problem 2.5 The line of equation $y=-x+1$ does descend. The line of equation $y=3x-2$ does rise.

Problem 2.6 Multiplying both sides of the equation $2x-3y+4=0$ by 2, we obtain that $4x-6y+8=0$ and then move the constant term to the right to obtain $4x-6y=-8$, which implies that two lines coincide.

Problem 2.7 Slope$=-2$, y-intercept$=4$, x-intercept$=2$.

Problem 2.8 Since $-2=a\times 0+b$ and $0=a\times 2+b$, it follows that $a=1$ and $b=-2$.

Problem 2.9 Since $2a-1-a^2+1=0$, it follows that $a=0$ or $a=2$.

Problem 2.10 y-intercept$=4$, x-intercept$=3$.

Problem 2.11 Slope of $AB=$Slope of $BC=3$.

Problem 2.12 The equation of the line AB is $y=-x+8$. Since C belongs to AB, $a=-a+8$, so $a=4$.

Problem 2.13 The midpoint is $M(2,1)$. The equation is $y=\frac{1}{2}x$.

Problem 2.14 $(-3,1)$, $(1,3)$, $(7,-5)$.

Problem 2.15 The first two lines meet at $A(1,-1)$. Since A also belongs to the third line, $1-1+a=0$, so $a=0$.

Problem 2.16 Dividing both sides of the equation $6x+8y+12=0$ by 2, we obtain that $3x+4y+6=0$. The distance is $\frac{|1-6|}{\sqrt{3^2+4^2}}=1$.

Problem 2.17 (1) $x\geqslant\frac{4}{3}$. (2) $y\leqslant 2$.

Problem 2.18 (1) $x>\frac{5}{2}$. (2) $x\geqslant 4$.

Problem 2.19 (1) $x\in(-\infty,-2)\cup(-1,3)$. (2) $x\in(-\infty,-2]\cup[1,2]$.

Problem 2.20 (1) $x\in(-3,2]$. (2) $x\in(-\frac{1}{3},4)$.

Problem 2.21 (1) $x\in(-2,0)$. (2) $x\in[1,3]$.

Problem 2.22 We may point out the results together with the graphs. Areas below the straight lines $y-2x=6$ and $2y+x=6$, respectively (the graph is omitted).

Problem 2.23 (1) $k=\pm 3$. (2) $k=0,4$.

Problem 2.24 $x^2+y^2+8x-2y+13=0$.

Problem 2.25 $(x-2)^2+(y-1)^2=5^2$.

Chapter 3 Functions

3.1 Mappings and functions

3.1.1 Relations and mappings

In everyday life, we encounter a large variety of relationships. In the most general sense, a relation is simply a correspondence between two sets.

Definition 3.1.1—Relation. A **relation** is a correspondence between two sets where each element in the first set, called the **domain**, corresponds to at least one element in the second set, called the **codomain.**

Relations can be represented in many different ways and may even be very "unmathematical". For instance,

(a) Every person has a blood type,

(b) Every person has a birthday,

as shown in Figure 3.1. The Cartesian product of two sets X and Y is the set of all ordered pairs, written (x,y), where x is an element of X and y is an element of Y. Then x and y are called the **components** of the ordered pair. A relation can also be expressed as a set of ordered pairs. For example, the relation in Figure 3.1(a) can be written as

{(Michael,A), (Mary,O), (Frank,AB), (Juliet,O), (Daniel, A)}.

The domain is the set of all the first components of the ordered pairs, and the **range** (which may not always be the same as the codomain) is the set of all the second components of the ordered pairs. As shown in Figure 3.1(b), X represents a set of people and Y represents a set of birthdays, we say that elements of X correspond to elements of Y, or the birthday relation maps elements of X to elements of Y.

Notation 3.1 Using what is called **mapping notation**, we might simply write a relation between X and Y as $X \to Y$.

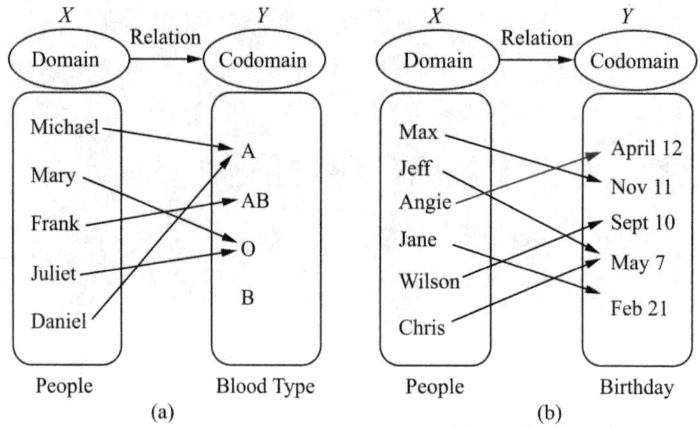

Figure 3. 1 Two relations illustrated by Venn diagrams.

Ⓡ From a practical standpoint, we note that while it is possible for two different people to share the same birthday, it is quite impossible for the same person to have two different birthdays. This observation will help us mark the difference between a relation and a special kind of relation called a **function.** A relation in which each element in the domain corresponds to exactly one element in the range is a function (see Figure 3. 2).

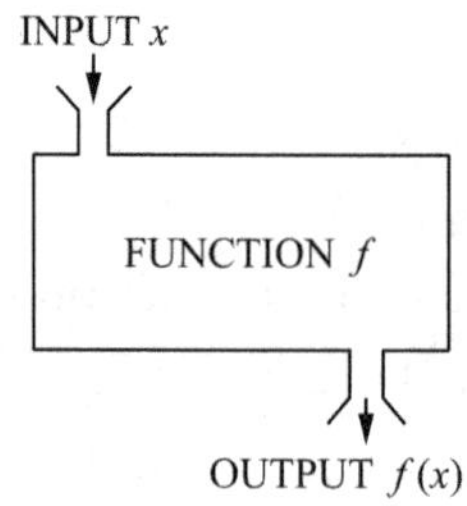

Figure 3. 2 A function f takes an input x and returns a single output $f(x)$. In metaphorical terms, a function is a "machine" or a "black box" that for each input returns a corresponding output.

3.1.2 Functions

Definition 3. 1. 2—Function. A function f from X to Y is a subset of the Cartesian product $X \times Y$ subject to the following condition: every element of X is the first component of one and only one ordered pair in the subset.

In other words, for every x (called the **argument** or the **independent variable**) in X there is exactly one element y (called the **image** of x, or the **dependent variable**, since it depends on x) such that the ordered pair (x, y) is contained in the subset defining the function f.

Ⓡ The definition of a function is more restrictive than the definition of a relation.

Ⓡ A couple of important and often overlooked facts about functions are the following:

(1) The range of a function depends upon the domain of the function.

(2) Functions do not have to be numerically valued or have numerical inputs.

As we know, for a relation, each input corresponds to **at least one** output, whereas for a function, each input corresponds to **exactly one** output. The birthday example given is both a relation and a function. Also note that the range (set of values to which the elements of the domain correspond) is a subset of the set of days in a year.

The difference between relations and functions

Although all functions are relations, not all relations are functions. For instance, suppose that the schedule of a university sporting event with five primary sports: Football, Volleyball, Basketball, Swimming and Tennis on a given day is represented by the relation shown in Figure 3.3(a). Because an element in the schedule, 2:00 PM, corresponds to **more than one** sport (two in this case, Football and Swimming), the correspondence between starting times and sports made precisely in the schedule below is **not a function.** However, it is a relation.

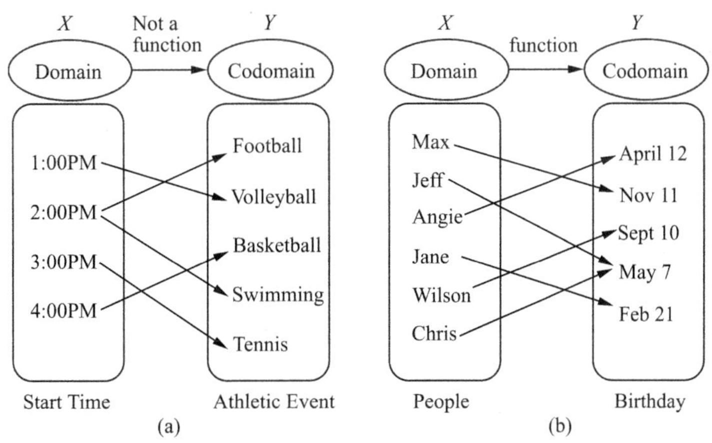

Figure 3.3 Two relations illustrated by Venn diagrams.

Definition 3.1.3—Domain. The domain of a function is the complete set of all possible values of the independent variable.

Definition 3.1.4—Codomain. The codomain or target set of a function is the set into which all outputs of the function are constrained to fall.

■ **Example 3.1** For the function defined by Figure 3.3(b), the domain is the set

{Max, Jeff, Angie, Jane, Wilson, Chris}

and the codomain is the set

{April 12, Nov 11, Sept 10, May 7, Feb 21}.

■

® A function whose domain and codomain are both subsets of the set of real numbers is called a **numeric function.**

Notation 3.2 A function f is commonly declared by stating its domain X and codomain Y, using the notation $f: X \to Y$ or $X \xrightarrow{f} Y$. The **functional notation** is the notation for expressing the image corresponding to the argument x as $f(x)$, first used by Leonhard

Euler in 1734.

Definition 3.1.5—Implied domain. When a numeric function f is defined, it is often the case that the domain of f is not explicitly stated. In such a case, we shall consider the domain to be the set of real numbers x so that the value of $f(x)$ is also a real number. This is called the **implied domain** (or the maximal domain) of f.

■ **Example 3.2** For instance, the function given by

$$f(x)=\frac{1}{x^2-9}$$

has an implied domain that consists of all real numbers other than $x=\pm 3$. These two values are excluded from the domain because substituting the values in the function will make the denominator by zero and division by zero is **undefined**. The set $(-\infty,-3)\cup(-3,3)\cup(3,\infty)$ is the implied domain of the function.

■

Definition 3.1.6—Range. The **range** of a function is the complete set of all possible resulting values of the dependent variable (y, usually), after we have substituted the argument of the function with all possible values in the domain.

® The codomain and range are both on the output side, but are subtly different. The codomain is the set of values that could possibly come out. The codomain is actually part of the definition of the function. The range is the set of values that actually do come out, it is not a part of the definition of the function and it is a subset of the codomain.

® **Characteristics of a function from a set X to a set Y:**

- Each element in X must be matched with an element in Y.
- Some elements in Y may not be matched with any element in X.
- Two or more elements in X may be matched with the same element in Y.
- An element in X (the domain) cannot be matched with two different elements in Y.

® It is important to note that:

- $f(x)$ does not mean f times x.
- The most common function names are f and F since the word function begins with "f". Other common function names are g and G, but any letter can be used.
- The letter most commonly used for the independent variable is x. The letter t is also common because it represents time in real-world applications, but any letter can be used. For instance,

$$f(x)=x^3+5x-2,$$
$$g(t)=t^3+5t-2,$$
$$h(s)=s^3+5s-2,$$

all define the same function.

- Although we can think of y and $f(x)$ as interchangeable, the function notation is useful when we want to consider two or more functions of the same independent variable or when we want to evaluate a function at more than one argument.

■ **Example 3.3** As shown in Figure 3.4, the set A is the domain, the set B is the codomain, and the set of elements that get pointed to in B (the actual values produced by the function) is the range. Then we have

- domain: $\{1,2,3,4\}$;
- codomain: $\{1,2,3,4,5,6,7,8,9,10\}$;
- range: $\{3,5,7,9\}$.

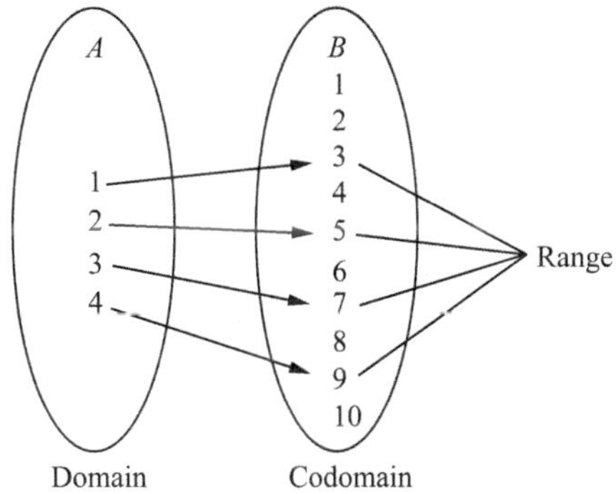

Figure 3.4 The function $x \to 2x-1$ with the domain $\{1,2,3,4\}$.

■

■ **Example 3.4** Find the implied domain of the following numeric functions:

(1) $f(x)=2x^3-5x^2$.

For this function, there is no value that will make it undefined. Therefore, the domain is the set of all real numbers, that is, $(-\infty,\infty)$.

(2) $g(x)=\dfrac{1}{x^2+4}$.

For this function, there is no real value that will make it undefined, since in finding the zeros of the denominator we have to solve the equation $x^2+4=0$, which has no real roots. Therefore, the domain is the set of all real numbers, that is, $(-\infty,\infty)$.

(3) $h(x)=\dfrac{\sqrt{x+2}}{x^2-9}$.

In the numerator (top) of this fraction, we have a square root. To make sure the values under the square root are non-negative, we can only choose x-values greater than or equal to -2 $(x\geqslant -2)$. The denominator (bottom) has x^2-9, which can be written as $(x+3)(x-3)$. So our values for x cannot include ± 3. We don't need to worry about the -3 anyway, because we decided in the first step that $x\geqslant -2$. So the domain for this function is $[-2,3)\cup(3,\infty)$.

(4) $q(x)=\dfrac{x-1}{\sqrt{x+4}}$.

First, the denominator cannot be zero. Also the value under the square root can not be negative. Therefore we use $x+4>0$ instead of $x+4\geqslant 0$ since the denominator cannot be zero, that is, $x>-4$. Thus, the domain contains all real numbers in the interval $(-4,\infty)$.

■

■ **Example 3.5** Find the range of the following functions:

(1) $f:\{-1, 0, 1\} \to \{0, 1, 2, 5\}$, $f(x)=x^2$.

Since the domain for this function is $\{-1, 0, 1\}$, when we substitute all values of x in this set, the resulting values of $f(x)$ are 0 and 1. Thus, the range is the set $\{0, 1\}$.

(2) $g:[1, 3] \to (-\infty, \infty)$, $g(x)= 3x-2$.

Since the domain for this function is $[1, 3]$, it is seen that

$1 \leqslant x \leqslant 3 \Rightarrow 3 \leqslant 3x \leqslant 9$ (multiplication by 3) $\Rightarrow 1 \leqslant 3x-2 \leqslant 7$ (subtraction of 2).

Thus, the range is the interval $[1, 7]$.

(3) $h:(-\infty, \infty) \to (-\infty, \infty)$, $g(x)=x^2+4$.

Since the domain for this function is $(-\infty, \infty)$, when we substitute all values in this interval, the resulting values of $g(x)$ will be at least 4 (since $x^2 \geqslant 0$), but may become arbitrarily large (since x^2 may become arbitrarily large). Thus the range is the interval $[4, \infty)$.

■

Exercise 3.1 Determine whether the following relations are functions:

(1) $\{(-1, 4), (2, 4), (1, 5), (3, 4)\}$.

(2) $\{(0, 4), (1, 3), (0, 5), (2, 2)\}$.

(3) *Domain*=Set of all items for sale in a grocery store and *Range*=Price.

■

Solution

(1) No x-value is repeated. Therefore, to each x-value corresponds exactly one y-value. This relation is a function.

(2) Both $y=4$ and $y=5$ correspond to the value $x=0$. This relation is not a function.

(3) To each item in the grocery store corresponds exactly one price. This relation is a function.

■

■ **Example 3.6** What is the range of $f(x)= x^2$ if its domain is $[2, \infty)$?

The range of $f(x)= x^2$ is $[0, \infty)$ if no constraints are given on x. In this case, though, we are given constraints on the domain, so it is likely that constraints will also exist on the range. Since $x \geqslant 2$ and is always positive, it follows that $x \cdot x \geqslant 2 \cdot 2=4$, so the range is $[4, \infty)$.

■

■ **Example 3.7** Sketch the graph of $f(x)= x^2$.

For this function, as shown in Figure 3.5, we are plotting all points whose y-coordinate is the square of the x-coordinate.

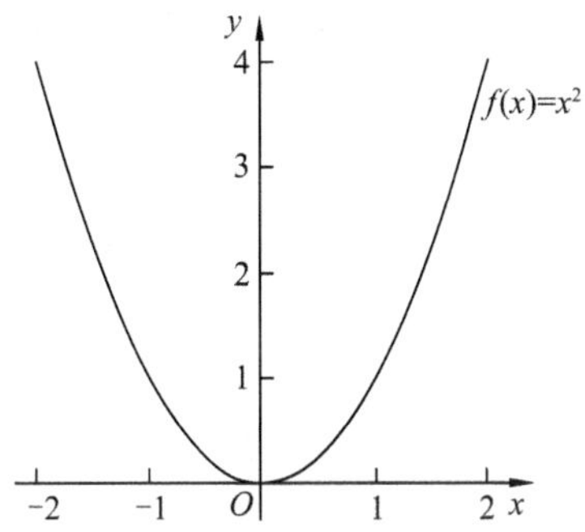

Figure 3.5 The function $f(x)=x^2$

■

■ **Example 3.8** Give an example of function modeling a real life situation where the algebraic domain of the function in the formula is different to the real life constraints.

Consider the formula for the volume of a sphere of radius r, $V=\frac{4}{3}\pi r^3$. The algebraic domain for this function is the set of all real numbers. However, a negative value of r, the radius, does not make sense in this context, so the physical domain is $r>0$, or the interval $(0,+\infty)$.

■

3.1.3 Four ways to represent a function

- **Verbally**, through a sentence that describes how the input variable is related to the output variable.
- **Numerically**, through a table or a list of ordered pairs that matches input values with output values.
- **Graphically**, through points on a graph in a coordinate plane in which the input values are represented by the horizontal axis and the output values are represented by the vertical axis.
- **Algebraically**, through a function described via some specific formula.

In traditional calculus and pre-calculus courses, most functions considered are represented with formulas and graphs and only occasionally they are given numerically. This leads to the obvious question of why we should consider all these different ways of describing functions. The answer to this lies in how scientific discovery progresses: as scientists trying to model a given phenomenon, we would ideally like some algebraic expression and in order get an algebraic formula, there are a number of different steps usually taken which use all these different representations:

(1) Identify and verbalize the problem.

(2) Run experiments and collect and write up data.

(3) Use data to sketch a graph.

(4) Use graph to find a formula.

■

■ **Example 3.9** Is the color of a car a function of its model? Why or why not?

If the color of a car is a function of model, that would mean when we input a model into the function, it would output a color. The only way this could be a function is if the car manufacturer made that car only in a single color, which is definitely not the case (nowadays, even the Ferraris are in other colors than red as well). Thus, this is not a function.

■

■ **Example 3.10** Which of the following define functions?

(1) $y=x^2-x$.

This certainly is a function if we define the independent variable to be x and the dependent variable to be y. However, nothing has been stated in the question to suggest that is the case, and it is clear that if x is the dependent variable, then this does not define a function. Thus, this defines y as a function of x, but not x as a function of y.

(2) $y^2=x-1$.

This defines x as a function of y, but not y as a function of x.

(3) The graph shown in Figure 3.6.

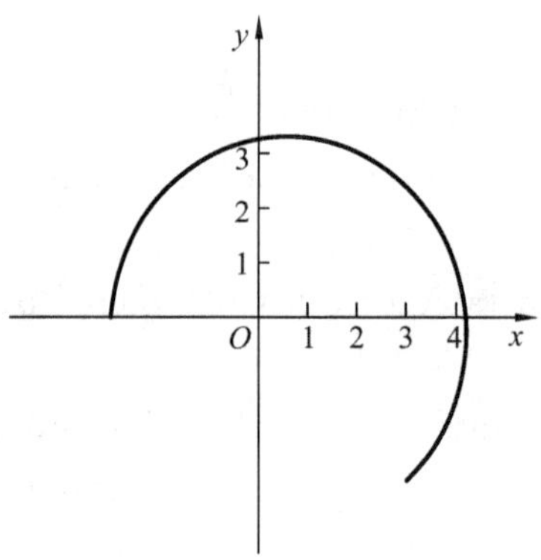

Figure 3.6 A circular arc.

This cannot possibly define a function since $x=4$ has two corresponding y values, and by definition of a function, any given input has a unique output. Similarly, $y=4$ has two corresponding x values, and consequently the graph does not define neither y as a function of x, nor x as a function of y.

■

■ **Example 3.11** Given the function $f(x)=2x^2-3x+1$, find the values of $f(1)$, $f(-2)$, $f(0)$ and $f(a+1)$.

(1) For $x=1$, $f(1)=2(1)^2-3(1)+1=0$.

(2) For $x=-2$, $f(-2)=2(-2)^2-3(-2)+1=15$.

(3) For $x=0$, $f(0)=2(0)^2-3(0)+1=1$.

(4) For $x=a+1$, $f(a+1)=2(a+1)^2-3(a+1)+1=2a^2+4a+2-3a-3+1=2a^2+a$.

■

3.1.4 Injective, surjective and bijective functions

Definition 3.1.7—Injective function. In mathematics, an **injective function** or **injection** or **one-to-one function** is a function that preserves distinctness: it never maps distinct elements of its domain to the same element of its codomain. In other words, every element of the function's range is the image of at most one element of its domain.

Notation 3.3 Occasionally, an injective function from X to Y is denoted by an arrow with a tail, as in $f: X \rightarrowtail Y$.

Ⓡ Symbolically, if $f: X \to Y$, then f is said to be injective provided that

$$\forall a, b \in X, a \neq b \Rightarrow f(a) \neq f(b),$$

or, alternatively,

$$\forall a, b \in X,\ f(a) = f(b) \Rightarrow a = b.$$

Ⓡ An injective function does not repeat values. That is, if there are $a, b \in X$, $a \neq b$, such that $f(a) = f(b)$, that is, f takes the same value (at least) twice, then f is not injective.

■ **Example 3.12** The function $f: \mathbf{R} \to \mathbf{R}$ defined by $f(x) = 2x + 1$ is injective, since

$$\forall a, b \in \mathbf{R},\ f(a) = f(b) \Rightarrow 2a + 1 = 2b + 1 \Rightarrow 2a = 2b \Rightarrow a = b.$$

■

■ **Example 3.13** The function $g: \mathbf{R} \to \mathbf{R}$ defined by $g(x) = x^2$ is not injective, because $g(1) = 1 = g(-1)$. However, if g is redefined so that its domain is the set of non-negative real numbers $[0, +\infty)$, then g is injective, since

$\forall a, b \in [0, \infty)$, $g(a) = g(b) \Rightarrow a^2 = b^2 \Rightarrow a = b$, since a and b are both non-negative.

■

Definition 3.1.8—Surjective function. In mathematics, a function f from a set X to a set Y is **surjective** (or **onto**), or a **surjection**, if for every element y in the codomain Y of f there is at least one element x in the domain X of f such that $f(x) = y$. It is not required that x is unique; the function f may map one or more elements of X to the same element of Y.

Notation 3.4 Surjections are sometimes denoted by a two-headed rightwards arrow, as in $f: X \twoheadrightarrow Y$.

Ⓡ Symbolically, if $f: X \to Y$, then f is said to be surjective if

$$\forall y \in Y, \exists x \in X \text{ such that } f(x) = y.$$

Ⓡ A surjective function takes all values in its codomain. That is, if there is $y \in Y$ such that $f(x) \neq y$ for any $x \in X$ (the value y is not taken), then f is not surjective.

Definition 3.1.9—Bijective function. In mathematics, a **bijection**, **bijective function** or **one-to-one correspondence** is a function between the elements of two sets, where each element of one set is paired with exactly one element of the other set, and each element of the other set is paired with exactly one element of the first set.

Notation 3.5 A two-headed rightwards arrow with a tail is sometimes used as a symbol to denote bijections.

Ⓡ The term one-to-one function must not be confused with one-to-one correspondence (bijective function), which uniquely maps all elements in both domain and codomain to each other.

Ⓡ A bijective function $f: X \to Y$ is a one-to-one (injective) and onto (surjective) mapping of a set X into a set Y.

■ **Example 3.14** The function shown in Figure 3.7 is injective, but non-surjective.

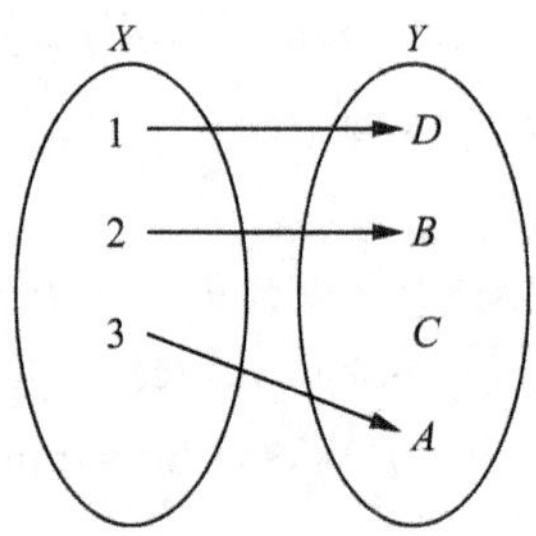

Figure 3.7 An injective, but non-surjective function (injection, not a bijection).

The function shown in Figure 3.8 is non-injective and surjective.

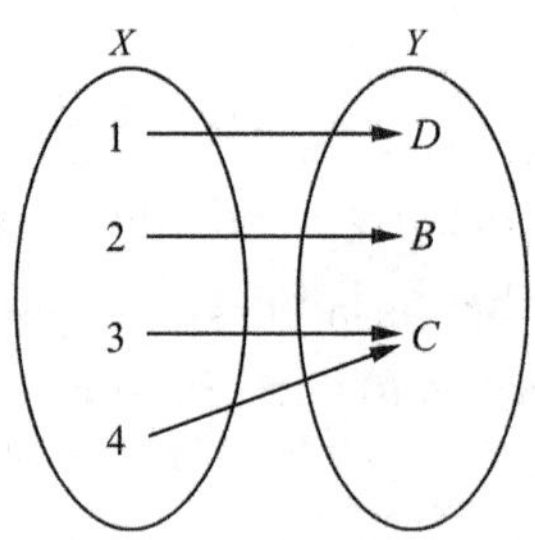

Figure 3.8 A non-injective surjective function (surjection, not a bijection).

The function shown in Figure 3.9 is bijective, since it is both injective and surjective.

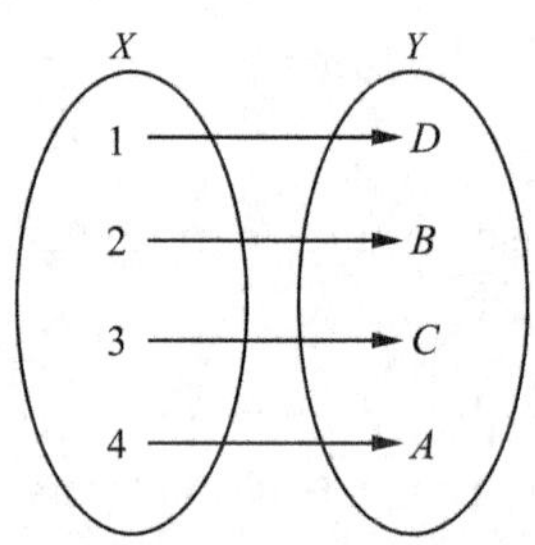

Figure 3.9 An injective and surjective function (bijection).

■

■ **Example 3.15** The function $f: \mathbf{R} \to \mathbf{R}$ defined by $f(x) = x^3 - 3x$ is surjective, because the pre-image of any real number y is the solution set of the cubic polynomial equation $x^3 - 3x - y = 0$ and every cubic polynomial with real coefficients has at least one real root.

However, this function is not injective (and hence not bijective). This happens since

$f(x) = x^2(x-3)$ and $f(0) = f(3) = 0$.

■

■ **Example 3.16** Any linear function over the set of reals, $f: \mathbf{R} \to \mathbf{R}$, $f(x) = ax+b$ (a is non-zero) is a bijection. Each real number y is uniquely obtained from the real number $x = \frac{y-a}{b}$.

■

3.1.5 Piecewise-defined functions

Sometimes, a single formula is not enough to describe a function and we may require a number of different formulas which depend upon which part of the domain an input comes from.

Definition 3.1.10—Piecewise-defined function. In mathematics, a **piecewise-defined function**, also called a **piecewise function** or a **hybrid function**, is a function which is defined by multiple sub-functions, each sub-function applying to a certain interval of the main function's domain.

® A piecewise function is a function that is defined on a sequence of intervals.

We illustrate with several examples.

■ **Example 3.17** A common example is the absolute value,

$$|x| = \begin{cases} x, & x>0, \\ 0, & x=0, \\ -x, & x<0. \end{cases}$$

■

■ **Example 3.18** A company charges a flat rate of 1 000 to rent up to 100 desks and every additional desk costs 8. Find a formula for the cost C as a function of n, the number of desks rented.

We need to break the domain up into two pieces where the function has different formulas-when 100 or less desks are rented, the function is constant with value 1 000 and when more than 100 are rented, it will be 1 000 plus 8 times any additional desks. Thus we get

$$C(n) = \begin{cases} 1\,000, & 0<n\leqslant 100, \\ 1\,000+8(n-100), & n>100,\ n \in \mathbf{Z}_+. \end{cases}$$

■

Exercise 3.2 Evaluate the function when $x=-1, 0$, and 1.

$$f(x) = \begin{cases} x^2+1, & x<0, \\ 2x-1, & x\geqslant 0. \end{cases}$$

■

Solution Because $x=-1$ is less than 0, we use the expression $f(x) = x^2+1$ to obtain $f(-1) = (-1)^2+1=2$. For $x=0$, we use the expression $f(x) = 2x-1$ (because this is the branch where the equal sign is) to obtain $f(0) = 2 \cdot 0-1=-1$. For $x=1$, which is greater than 0, we use the expression $f(x) = 2x-1$ to obtain $f(1) = 2 \cdot 1-1=1$.

■

3.2 Inverse functions

Definition 3.2.1—Inverse function. In mathematics, an inverse function is a function that "reverses" another function: if the function f applied to an input x gives a result of y, then applying its inverse function g to y gives the result x, and vice versa.

® If f is invertible, then the function g is unique.

® If f is invertible, it then must be bijective.

Notation 3.6 The inverse function g is usually denoted as f^{-1}.

® The involutory nature of the inverse can be concisely expressed by $(f^{-1})^{-1}=f$.

For example, the reversing effect of the function $f(x) = x+4$ from the set $A=\{1,2,3,4\}$ to set $B=\{5,6,7,8\}$ is written as $f^{-1}(x) = x-4$ which is the inverse of the function $f(x)$. The range (set B) of $f(x)$ becomes the domain (set A) of $f^{-1}(x)$. For this example the inverse is a function (see Figure 3.10).

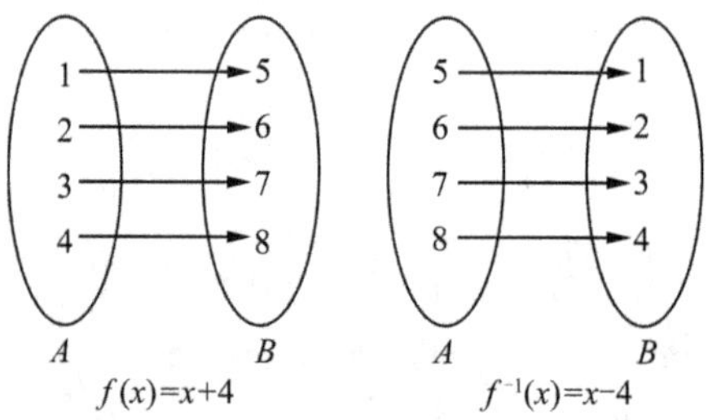

Figure 3.10 A function f and its inverse f^{-1}.

Now consider the function $f(x) = x^2$, defined for the domain $\{x \mid -2 \leqslant x \leqslant 2,\ x \in \mathbf{Z}\}$. The reverse of the function is $\pm\sqrt{x}$ which by definition is not a function (see Figure 3.11).

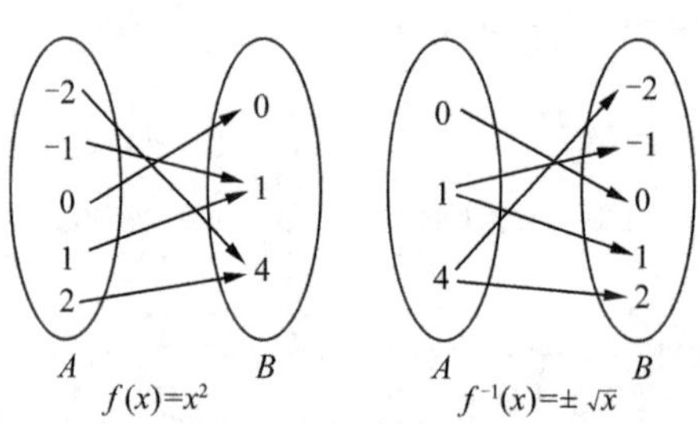

Figure 3.11 $f(x)=x^2, \{x \mid -2 \leqslant x \leqslant 2,\ x \in \mathbf{Z}\}$ **and its inverse is not a function.**

Note that only those functions which, for a given domain, are one-to-one mappings, have inverse functions. This point can be illustrated graphically by changing the domain $\{x \mid -2 \leqslant x \leqslant 2,\ x \in \mathbf{Z}\}$ over which the function $f(x) = x^2$ is defined. It is possible to define the inverse as a function if we restrict the domain to values of x for which the mapping is one-to-one. For example, if $f(x)=x^2$ is defined for the restricted domain $\{x \mid 0 \leqslant x \leqslant 2\}$, then the inverse function $f^{-1}(x)=\sqrt{x}$ does exist (see Figure 3.12).

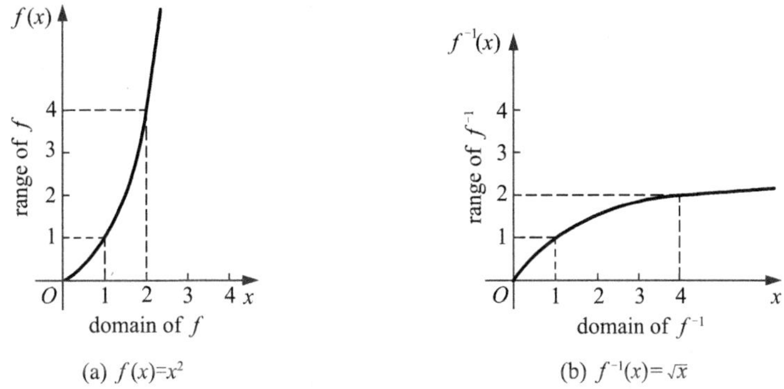

Figure 3.12 $f(x)=x^2$, $\{x \mid 0 \leqslant x \leqslant 2\}$ **and its inverse is a function.**

3.2.1 How to find the inverse of a function

To find the inverse of a function, express the function in the form $y=f(x)$, interchange x and y, and then rearrange the equation to make y the subject of the equation again. If f is an invertible function with domain X and range Y, then its inverse f^{-1} has domain Y and range X (see Figure 3.13).

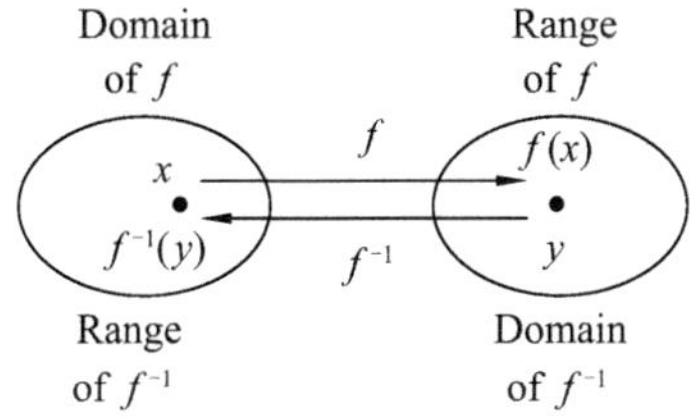

Figure 3.13 $f(x)$ **and its inverse** $f^{-1}(x)$: **if** f **is an invertible function with domain** X **and range** Y, **then its inverse** f^{-1} **has domain** Y **and range** X.

Exercise 3.3 The function $g(x)=\dfrac{x+1}{x-3}$ is defined for the domain $\{x \mid x \neq 3, x \in \mathbf{R}\}$. Find its inverse function, and state the value of x for which $g^{-1}(x)$ is not defined. ■

Solution Writing the function in the form $y=g(x)$, and interchanging x and y gives $x=\dfrac{y+1}{y-3}$. Rearrange the equation to make y the subject.

$$
\begin{aligned}
& x(y-3)=y+1 \\
\Rightarrow\quad & xy-3x=y+1 \\
\Rightarrow\quad & xy-y=3x+1 \\
\Rightarrow\quad & y(x-1)=3x+1 \\
& y=\frac{3x+1}{x-1}.
\end{aligned}
$$

So,

$$g^{-1}(x)=\frac{3x+1}{x-1}.$$

Try putting $x=1$ into the inverse function. What happens? Division by zero causes a problem, creating an answer that is undefined. So this inverse function is defined for all real values of x except $x=1$.

■

3.2.2 How to graph the inverse of a function

The graphs of a function f and its inverse function f^{-1} are related to each other in the following way. If the point (a,b) lies on the graph of f, then the point (b,a) must lie on the graph of f^{-1} and vice versa. This means that the graph of f^{-1} is a reflection of the graph of f in the line $y=x$ (see Figure 3. 14).

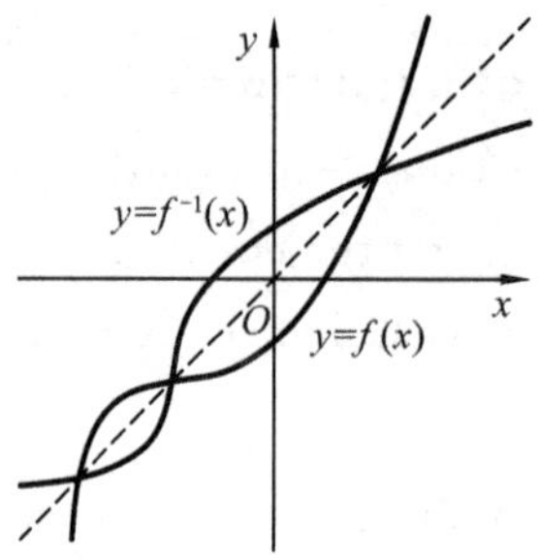

Figure 3. 14 The graphs of $f(x)$ and its inverse $f^{-1}(x)$. The dotted line is $y=x$.

■ **Example 3. 19** Sketch the graphs of the inverse functions $f(x)=x^2\ (x\geqslant 0)$ and $f^{-1}(x)=\sqrt{x}$ on the same rectangular coordinate system and show that the graphs are reflections of each other in the line $y=x$.

The graphs of f and f^{-1} are shown in Figure 3. 15. It appears that the graphs are reflections of each other in the line $y=x$. You can further verify this reflective property by testing a few points on each graph.

Graph of $f(x)=x^2$, $x\geqslant 0$	Graph of $f^{-1}(x)=\sqrt{x}$
(0,0)	(0,0)
(1,1)	(1,1)
(2,4)	(4,2)
(3,9)	(9,3)

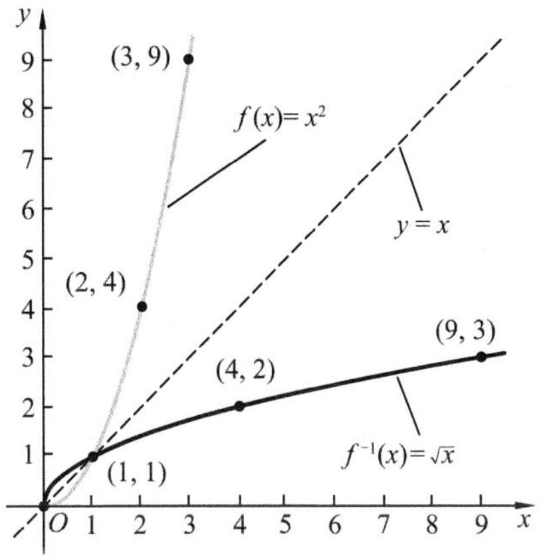

Figure 3.15 The square root of x is a partial inverse to $f(x)=x^2$. The dotted line is $y=x$.

■

3.3 Composite functions

3.3.1 Arithmetic combinations of functions

Just as two real numbers can be combined using arithmetic operations, the output values of two or more functions can be combined using addition, subtraction, multiplication and division. For instance, if $f(x)=7x$ and $g(x)=\cos x$, possible combinations include:

$f(x)+g(x)=7x+\cos x.$

$f(x)-g(x)=7x-\cos x.$

$f(x)\times g(x)=7x\cos x.$

$f(x)\div g(x)=\dfrac{7x}{\cos x}.$

Exercise 3.4 Given $f(x)=2x+1$ and $g(x)=x^2+2x-1$, find:

(1) $(f-g)(x)$;

(2) $(f+g)(x)$;

(3) $(f\div g)(x)$;

(4) $(fg)(x)$.

■

Solution We can obtain the following results by direct computation.

(1) $(f-g)(x)=f(x)-g(x)=(2x+1)-(x^2+2x-1)=-x^2+2.$

(2) $(f+g)(x)=f(x)+g(x)=(2x+1)+(x^2+2x-1)=x^2+4x.$

(3) $(f\div g)(x)=\left(\dfrac{f}{g}\right)(x)=\dfrac{f(x)}{g(x)}=\dfrac{2x+1}{x^2+2x-1}.$

(4) $(fg)(x)=f(x)\times g(x)=(2x+1)(x^2+2x-1)=2x^3+5x^2-1.$

■

3.3.2 Composition of functions

Another way of combining two functions is to form the **composition** of one with the other. In mathematics, function composition is the pointwise application of one function to the result of another to produce a third function. For instance, if $f(x)=x^2$ and $g(x)=x+1$, the composition of f with g is given by

$$f(g(x))=f(x+1)=(x+1)^2.$$

Notation 3.7 This composition is denoted as $f\circ g$ and reads as " f circle g" , or " f round g", or " f composed with g", " f after g", " f following g", or " f of g", or" f on g".

Definition 3.3.1—Composition of two functions. The composition of the function f with function g is

$$(f\circ g)(x)=f(g(x)).$$

The domain of $f\circ g$ is the set of all x in the domain of g such that $g(x)$ is in the domain of f .

Exercise 3.5 Given $f(x)= x+2$ and $g(x)= 4-x^2$, find:

(1) $(f\circ g)(x)$;

(2) $(g\circ f)(x)$;

(3) $(g\circ f)(-2)$.

■

Solution

(1) $(f\circ g)(x)=f(g(x))=f(4-x^2)=(4-x^2)+2=-x^2+6$.

(2) $(g\circ f)(x)=g(f(x))=g(x+2)=4-(x+2)^2=-x^2-4x$.

(3) $(g\circ f)(-2)=-(-2)^2-4(-2)=-4+8=4$.

■

Ⓡ Note that $f\circ g$ is not the same as $g\circ f$. For function composition, the order matters.

Exercise 3.6 The function f,g and h are defined by

$$f:x\to\sqrt{x-1},\ g:x\to x^2+1,\ h:x\to 2x+3.$$

Find expressions for the following composite functions:

(1) $(g\circ f)(x)$;

(2) $(h\circ g)(x)$;

(3) $(g\circ h)(x)$;

(4) $(h\circ f)(x)$;

(5) $(h\circ (g\circ f))(x)$;

(6) $(f\circ(h\circ g))(x)$.

■

Solution

(1) $(g\circ f)(x)=g(f(x))=g(\sqrt{x-1})=(\sqrt{x-1})^2+1=x$.

(2) $(h\circ g)(x)=h(g(x))=2(x^2+1)+3=2x^2+5$.

(3) $(g\circ h)(x)=g(h(x))=(2x+3)^2+1=4x^2+12x+10$.

(4) $(h\circ f)(x)=h(f(x))=2\sqrt{x-1}+3$.

(5) $(h\circ(g\circ f))(x)=h(g(f(x)))=h(x)=2x+3$.

(6) $(f\circ(h\circ g))(x)=f(h(g(x)))=f(2x^2+5)=\sqrt{2x^2+4}$.

■

■ **Example 3.20** Write the function given by $h(x)=\frac{1}{(x-2)^2}$ as a composition of two functions.

In fact, one way to write h as a composition of two functions is to take the inner function to be $g(x)=x-2$ and the outer function to be $f(x)=\frac{1}{x^2}$. Then we can write

$$h(x)=\frac{1}{(x-2)^2}=f(x-2)=f(g(x))=f\circ g.$$

Therefore $h(x)=f\circ g$, where $f(x)=\frac{1}{x^2}$ and $g(x)=x-2$.

■

The domain of the composition of two functions f and g is illustrated in Figure 3.16. How is the domain of a composite function found? First, establish which values of x in the domain of function f produce values of $f(x)$ that lie in the domain of function g. These can then be fed into $g(x)$. However, depending upon the nature of functions f and g, and the values of x in the domains for which they are defined, the composite function $f\circ g(x)$ may have a restricted domain.

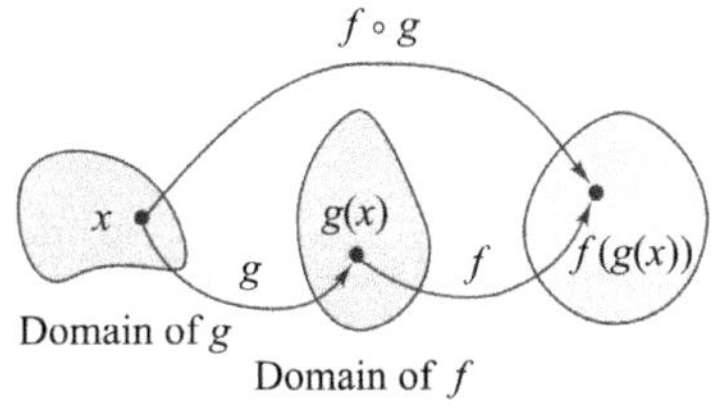

Figure 3.16 Domain of the composition of two functions f and g.

■ **Example 3.21** Given the function $f(x)=\frac{1}{x-1}$ and $g(x)=\frac{1}{x}$, determine $f\circ g$, and state its domain. Express the composition function $f\circ g(x)$ as

$$f(g(x))=\frac{1}{g(x)-1}=\frac{1}{\frac{1}{x}-1}=\frac{x}{1-x}.$$

For the domain of $f\circ g(x)$, from the final result of $f(g(x))$, we see that the denominator is zero when $x=1$. Therefore, $x\neq 1$. Are there any values of x that are not allowed? The function $g(x)$ has the domain $x\neq 0$; therefore, we must also exclude zero.

The domain of $f\circ g(x)=f(g(x))$ should exclude $x=0$ and $x=1$ or, in interval notation

$$(-\infty,0)\cup(0,1)\cup(1,\infty).$$

■

Exercise 3.7 The functions f and g are defined by

$$f(x)=1+2x \text{ for } \{x\mid x\geqslant 0, x\in\mathbf{R}\},$$

$$g(x)=\frac{1}{x-1} \text{ for } \{x\mid x>1, x\in\mathbf{R}\}.$$

(1) State the ranges of f and g for the given domains.

(2) Find an expression for $g\circ f$, and determine its domain.

(3) Find an expression for $f\circ g$, and determine its domain.

■

Solution

(1) For the given domain, the least value of $f(x)$ is 1 when $x=0$. Therefore, the range of $f(x)$ is $\{y\mid y\geqslant 1, y\in\mathbf{R}\}$ or $[1,+\infty)$. What about $g(x)$? As x increases, $x-1$ also increases. This means that as x increases $\frac{1}{x-1}$ decreases. As x becomes very large, $g(x)$ will tend toward zero. Conversely, for values of x close to 1, the denominator is very small and $g(x)$ becomes very large. This means that the range of $g(x)$ is $\{y\mid y>0, y\in\mathbf{R}\}$ or $(0,+\infty)$.

(2) $(g\circ f)(x)=g(1+2x)=\frac{1}{(1+2x)-1}=\frac{1}{2x}$. The domain of the composite function of $g\circ f(x)$ is the set of values of x in the domain of function f that give values of $f(x)$ that are in the domain of function g. Since the domain of function g is $\{x\mid x>1, x\in\mathbf{R}\}$, we need the values of x in the domain of f for which $f(x)>1$. Are these values in the domain of f that do not satisfy this? The only value of x in the domain of f that does not satisfy this condition is $x=0$. This means that the domain of the composite function $g\circ f$ is $\{x\mid x>0, x\in\mathbf{R}\}$ or $(0,\infty)$.

(3) $(f\circ g)(x)=f\left(\frac{1}{x-1}\right)=1+2\left(\frac{1}{x-1}\right)=\frac{x-1+2}{x-1}=\frac{x+1}{x-1}$. The domain of the composite function $f\circ g$ is the set of values of x in the domain of function g which give values of $g(x)$ that are in the domain of function f. Since the domain of function f is $\{x\mid x\geqslant 0, x\in\mathbf{R}\}$, we need the values of x in the domain of g for which $g(x)\geqslant 0$. So the domain of the composite function $f\circ g(x)$ is $\{x\mid x>1, x\in\mathbf{R}\}$.

■

The **inverse of this composite function** is $(f\circ g)^{-1}(x)$. This is obtained by reversing each of the individual component functions and the order is applied by

$$g^{-1}\circ f^{-1}(x)\overset{g^{-1}}{\longleftarrow}f^{-1}(x)\overset{f^{-1}}{\longleftarrow}x.$$

The values of x are put into f^{-1}, and then values of $f^{-1}(x)$ are put into g^{-1}. This gives

$$(f\circ g)^{-1}(x)\equiv g^{-1}\circ f^{-1}(x).$$

Ⓡ Note that $(f\circ g)^{-1}$ equals $g^{-1}\circ f^{-1}$, not $f^{-1}\circ g^{-1}$.

■ **Example 3.22** The functions f and g defined for $x\in \mathbf{R}$ by $f: x\to x^3$ and $g: x\to 2x+1$. Find:

(1) f^{-1} and g^{-1};

(2) $f^{-1}\circ g^{-1}$ and $(f\circ g)^{-1}$.

In fact, $f^{-1}(x)=\sqrt[3]{x}$ and $g^{-1}(x)=\frac{x-1}{2}$. $f^{-1}\circ g^{-1}=f^{-1}\left(\frac{x-1}{2}\right)=\sqrt[3]{\frac{x-1}{2}}$. Finally, one notes that

$$(f\circ g)^{-1}(x)=g^{-1}\circ f^{-1}(x)=g^{-1}(f^{-1}(x))=g^{-1}(\sqrt[3]{x})=\frac{\sqrt[3]{x}-1}{2}.$$

■

3.4 Transformation of graphs and functions

Definition 3.4.1—Parent function. In mathematics, a **parent** function is the simplest function of a family of functions that preserves the definition (or shape) of the entire family.

For example, for the family of quadratic functions having the general form $y=ax^2+bx+c$, the simplest function is $y=x^2$. This is therefore the parent function of the family of quadratic equations.

Graph shifting

Many functions have graphs that are simple transformations of the parent graphs. For example, you can obtain the graph of

$$h(x)=x^2+2$$

by shifting the graph of $f(x)=x^2$ **upward** two units as shown in Figure 3.17 (a) below. Similarly, you can obtain the graph of

$$g(x)=(x-2)^2$$

by shifting the graph of $f(x)=x^2$ to the **right** two units as shown in Figure 3.17 (b) below.

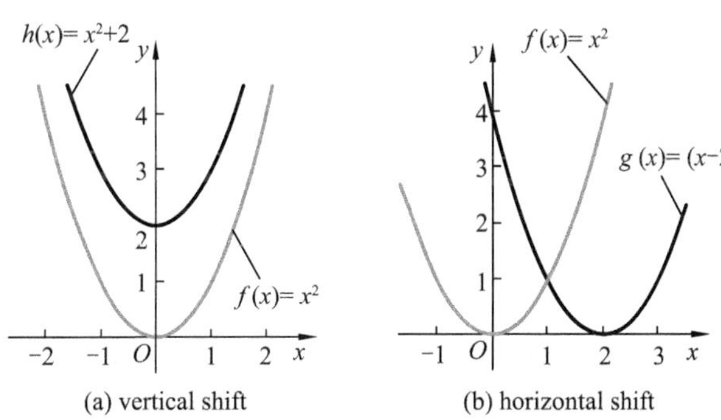

Figure 3.17 Shifting the graphs.

The following list summarizes this discussion about **horizontal and vertical shifts.**

(1) Vertical shifts c units *upward*: $h(x)=f(x)+c$.

(2) Vertical shifts c units *downward*: $h(x)=f(x)-c$.

(3) Horizontal shifts c units *to the right*: $h(x)=f(x-c)$.

(4) Horizontal shifts c units *to the left*: $h(x)=f(x+c)$.

■ **Example 3.23** Use the graph of $f(x)=x^3$ to sketch the graph of the following functions:

(1) $g(x)= x^3-1$;

(2) $h(x)=(x+2)^3+1$.

(1) Relative to the graph of $f(x)=x^3$, the graph of $g(x)=x^3-1$ is a downward shift of one unit, as shown in Figure 3.18 below.

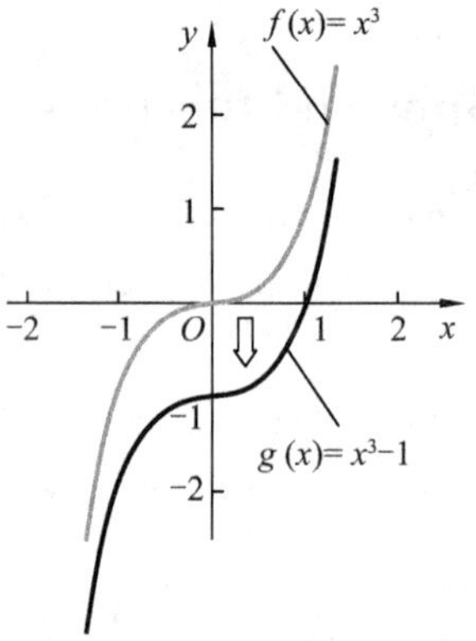

Figure 3.18 Shifting the graph.

(2) Relative to the graph of $f(x)=x^3$, the graph of $h(x)=(x+2)^3+1$ involves a left shift of two units and an upward shift of one unit as shown in Figure 3.19 below.

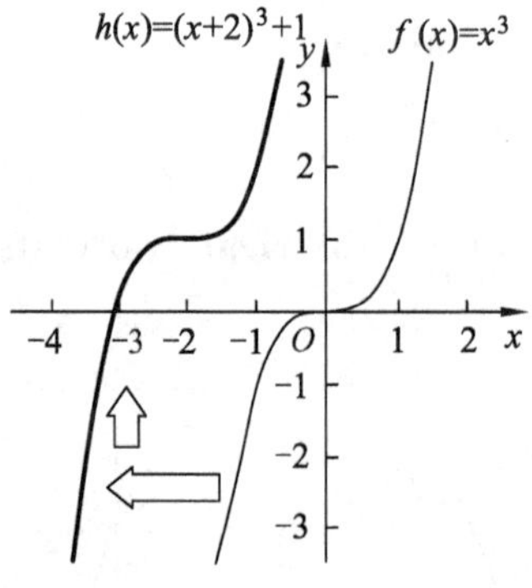

Figure 3.19 Shifting the graph.

■

Graph reflections

The second common type of transformation is a **reflection.** Reflections in the coordinate axes of the graph of $y=f(x)$ are represented as follows:

(1) Reflection in the x-axis, $h(x)=-f(x)$.

(2) Reflection in the y-axis, $h(x)=f(-x)$.

For instance, as shown in Figure 3.20, if you consider the x-axis to be a mirror, the graph of $h(x)=-x^2$ is the mirror image (or reflection) of the graph of $f(x)=x^2$.

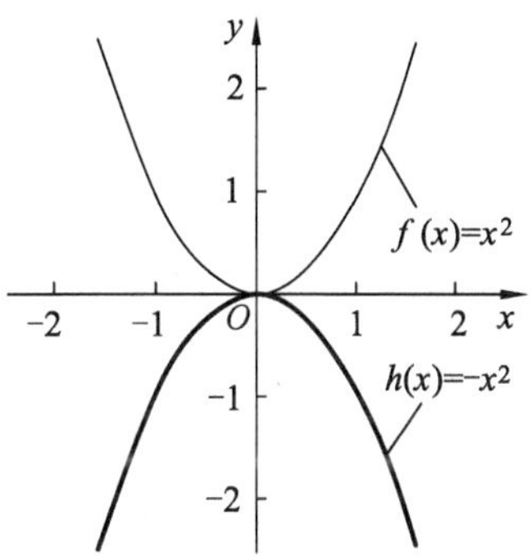

Figure 3.20 The graph of $h(x)=-x^2$ is the mirror image (or reflection) of the graph of $f(x)=x^2$.

■ **Example 3.24** Sketch the graph of each function by comparing it with the graph of $f(x)=\sqrt{x}$.

(1) $g(x)=-\sqrt{x}$;

(2) $h(x)=\sqrt{-x}$;

(3) $k(x)=-\sqrt{x+2}$.

In fact, as shown in in Figure 3.21, the graph of $g(x)$ is a reflection of the graph $f(x)$ in the x-axis, the graph of $h(x)$ is a reflection of the graph of $f(x)$ in the y-axis, and the graph of $k(x)$ is a left shift of two units of the graph of $f(x)$ followed by a reflection in the x-axis.

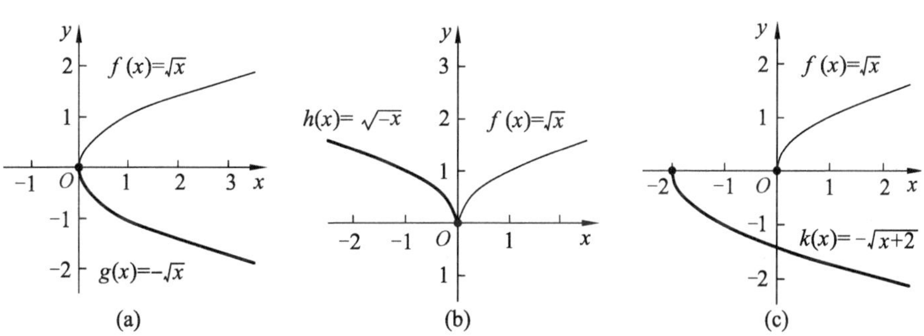

Figure 3.21 Reflecting graphs.

■

Graph shrinking and stretching

Horizontal shifts, vertical shifts, and reflections are **rigid transformations** because the basic shape of the graph is unchanged. These transformations change only the position of the graph in the coordinate plane.

Non-rigid transformations are those transformations that cause a distortion—a change in the shape of the original graph. For instance, a non-rigid transformation of the graph of $y=f(x)$ is represented by $g(x)=cf(x)$, where the transformation is a **vertical stretch** if

$c>1$ and a **vertical shrink** if $0<c<1$. Another non-rigid transformation of the graph of $y=f(x)$ is represented by $h(x)=f(cx)$ where the transformation is a **horizontal shrink** if $c>1$ and a **horizontal stretch** if $0<c<1$.

■ **Example 3.25** Compare the graph of each function with the graph of $f(x)$:

(1) $h(x)=3|x|$ and $g(x)=\frac{1}{3}|x|$ with $f(x)=|x|$.

(2) $g(x)=f(2x)$ and $h(x)=f\left(\frac{1}{2}x\right)$ with $f(x)$.

(1) Relative to the graph of the graph of $f(x)=|x|$, $h(x)=3|x|=3f(x)$ is a vertical stretch (each value is multiplied by 3) of the graph of $f(x)$. Similarly, the graph of $g(x)=\frac{1}{3}|x|$ is a vertical shrink (each value is multiplied by $\frac{1}{3}$) of the graph of $f(x)$.

(2) Relative to the graph of $f(x)$, the graph of $g(x)$ is a horizontal shrink ($c>1$) of the graph of f. Similarly, the graph of $h(x)=f\left(\frac{1}{2}x\right)$ is a horizontal stretch ($0<c<1$) of the graph of f.

■

■ **Example 3.26** Express the function $g(x)=2x^2+8x+6$ in the form $c[(x+a)^2+b]$, where a, b and c are constants. Describe the transformation by which the curve $y=x^2$ is mapped to the graph of this function. We shall complete the square of the quadratic function

$$g(x)=2(x^2+4x+3)=2(x^2+4x+4-1)=2[(x+2)^2]-1].$$

The graph of $y=x^2$ is mapped to the graph of $y=g(x)$ by a shift of 2 to the left and a shift of 1 unit downward, followed by a stretch of factor 2.

■

3.5 Odd, even and periodic functions

Graphs of certain classes of functions are symmetric in some way. We can categorise these functions as odd, even and periodic. A function is said to be **even** if its graph is symmetric with respect to the y-axis and to be **odd** if its graph is symmetric with respect to the **origin**.

Test for even and odd functions

- A function f is **even** if, for each x in the domain of f, $f(-x)=f(x)$.
- A function f is **odd** if, for each x in the domain of f, $f(-x)=-f(x)$.

■ **Example 3.27** The functions f, g and h are defined by $f(x)=x^3-2x$, $g(x)=1+x^2-5x^4$ and $h(x)=3x^2+2x$. Determine which of these functions are even or odd.

We check whether or not a function is even or odd by substituting $-x$ into the function in place of x.

$$f(-x)=(-x)^3-2(-x)=-x^3+2x=-(x^3-2x)=-f(x).$$

Therefore, f is an odd function.

$$g(-x)=1+(-x)^2-5(-x)^4=1+x^2-5x^4=g(x).$$

Therefore, g is an even function.

$$h(-x)=3(-x)^2+2(-x)=3x^2-2x.$$

Since $h(-x)\neq h(x)$ and $h(-x)\neq -h(x)$, the function h is neither even nor odd.

■

In mathematics, a periodic function is a function that repeats its values in regular intervals or periods.

Definition 3.5.1—Periodic function. A function f is said to be **periodic** with period P (P being a nonzero constant) if we have

$$f(x+P)=f(x)$$

for all values of x in the domain.

Definition 3.5.2—Fundamental period. If there exists a least positive constant P with the property that $f(x+P)=f(x)$ for all values of x in the domain, it is called the **fundamental period** (also **primitive period**, **basic period**, or **prime period**).

® If a function f is periodic with period P, then for all x in the domain of f and all positive integers n,

$$f(x+nP)=f(x).$$

® If $f(x)$ is a function with period P, then $f(ax)$, where a is a non-zero real number, is periodic with period $\frac{P}{|a|}$.

Geometrically, a periodic function can be defined as a function whose graph exhibits translational symmetry. Specifically, a function f is periodic with period P if the graph of f is invariant under translation in the x-direction by a distance of P. A simple example of a periodic function is the function f that gives the "fractional part" of its argument. Its period is 1. In particular,

$$f(0.5)=f(1.5)=f(2.5)=\cdots=0.5.$$

3.6 Exponentials

Definition 3.6.1—Exponential function. The exponential function f with base a is denoted by

$$f(x)=a^x,$$

where $a>0$, $a=1$ and x is any real number.

® The base $a=1$ is excluded because it yields $f(x)=1^x=1$. This is a constant function, not an exponential function.

3.6.1 Graphs of exponential functions

As functions of a real variable, exponential functions are uniquely characterized by the fact that the growth rate of such a function is directly proportional to the value of the function (see Figure 3.22).

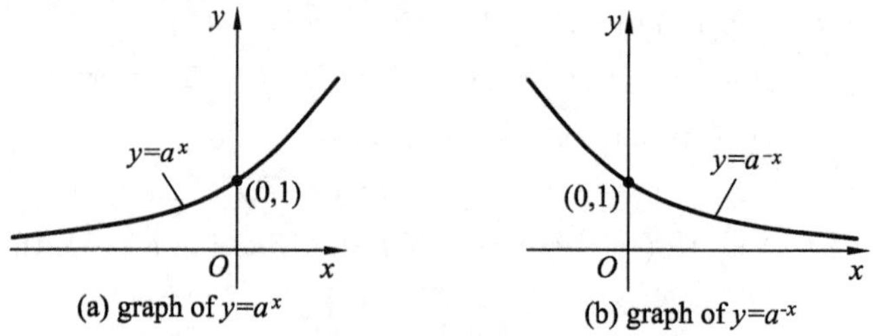

Figure 3.22 The graphs of a^x and a^{-x}.

Characteristics of graph of $y=a^x$, $a>1$

- Domain $(-\infty, \infty)$.
- Range $(0,\infty)$.
- y-intercept: $(0,1)$.
- Increasing.
- The graph gets close to the x-axis for x tending to $-\infty$.

Characteristics of graph of $y=a^{-x}$, $a>1$

- Domain $(-\infty, \infty)$.
- Range $(0,\infty)$.
- y-intercept: $(0,1)$.
- Decreasing.
- The graph gets close to the x-axis for x tending to ∞.

■ **Example 3.28** In the same coordinate plane, sketch the graph of each function:

(1) $f(x)= 2^x$.

(2) $g(x)=4^x$.

The table below lists some values for each function:

x	-3	-2	-1	0	1	2
2^x	$\frac{1}{8}$	$\frac{1}{4}$	$\frac{1}{2}$	1	2	4
4^x	$\frac{1}{64}$	$\frac{1}{16}$	$\frac{1}{4}$	1	4	16

Figure 3.23 shows the graphs of the two functions. Note that both graphs are increasing. Moreover, the graph of $g(x)=4^x$ is increasing more rapidly than the graph of $f(x)=2^x$

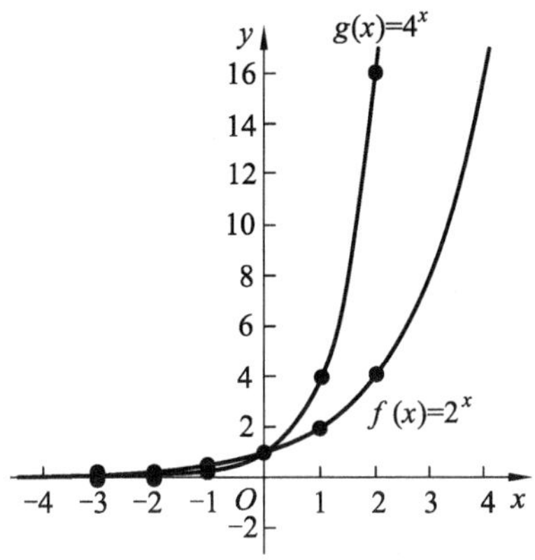

Figure 3. 23 The graphs of 2^x and 4^x.

■

■ **Example 3. 29** In the same coordinate plane, sketch the graph of each function:

(1) $F(x)= 2^{-x}$;

(2) $G(x)=4^{-x}$.

The table below lists some values for each function:

x	-2	-1	0	1	2	3
2^{-x}	4	2	1	$\frac{1}{2}$	$\frac{1}{4}$	$\frac{1}{8}$
4^{-x}	16	4	1	$\frac{1}{4}$	$\frac{1}{16}$	$\frac{1}{64}$

Figure 3. 24 shows the graphs of the two functions. Note that both graphs are decreasing. Moreover, the graph $G(x)=4^{-x}$ is decreasing more rapidly than the graph of $F(x)=2^{-x}$.

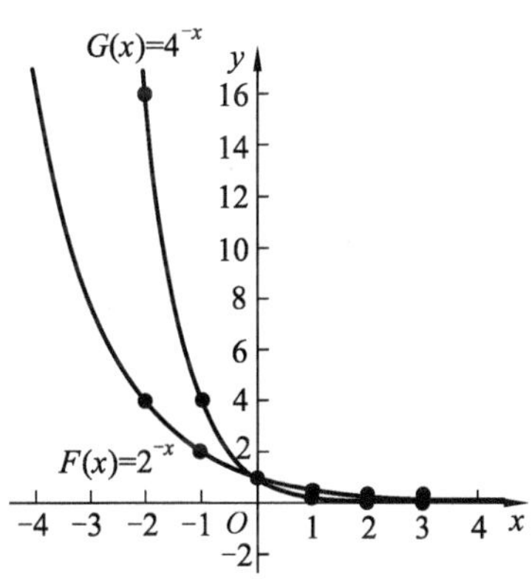

Figure 3. 24 The graphs of 2^{-x} and 4^{-x}.

■

3.6.2 Transformations of graphs of exponential functions

Let us use an example to illustrate the process. Each of the following graphs is a transformation of the graph of $f(x)=3^x$.

(1) Because $g(x)=3^{x+1}=f(x+1)$, the graph of g can be obtained by shifting the graph of f one unit to the *left* as shown in Figure 3. 25 (a).

(2) Because $h(x)=3^x-2=f(x)-2$, the graph of h can be obtained by shifting the graph of f two units *downward* as shown in Figure 3. 25 (b).

(3) Because $k(x)=-3^x=-f(x)$ the graph of k can be obtained by *reflecting* the graph of f in the x-axis, as shown in Figure 3. 25 (c).

(4) Because $j(x)=3^{-x}=f(-x)$ the graph of j can be obtained by *reflecting* the graph of f in the y-axis as shown in Figure 3. 25 (d).

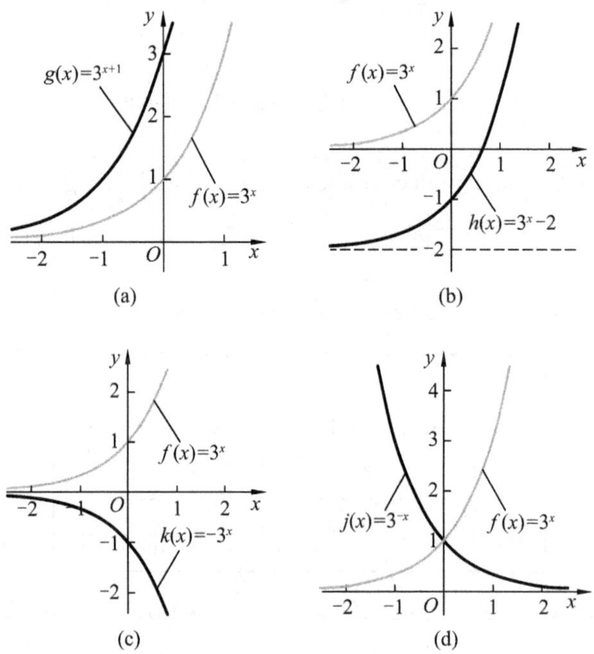

Figure 3. 25 The graphs of 3^{x+1}, 3^x-2, -3^x and 3^{-x}.

3.6.3 The natural base e

In many applications, the most convenient choice for a base is the irrational number

$$e=2.718\ 281\ 828\cdots.$$

This number is called the **natural base.** The function given by $f(x)=e^x$ is called the **natural exponential function.** Figure 3. 26 shows the graph of $f(x)=e^x$, where x is the variable.

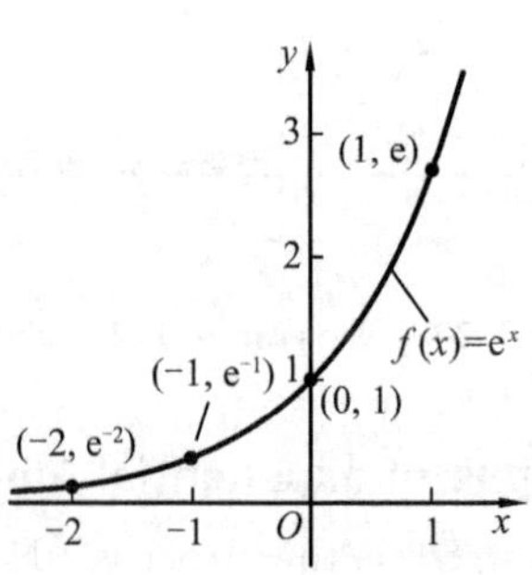

Figure 3. 26 The graph of e^x.

The exponential function as a series

The number e can also be defined as

$$e = 1 + \frac{1}{1!} + \frac{1}{2!} + \frac{1}{3!} + \cdots = \sum_{k=0}^{\infty} \frac{1}{k!}$$

In general, the function e^x which is often referred to as the **exponential function** can be defined as

$$e^x = 1 + \frac{x}{1!} + \frac{x^2}{2!} + \frac{x^3}{3!} = \cdots = \sum_{k=0}^{\infty} \frac{x^k}{k!}$$

■ **Example 3.30** Find the value of x in the following equations:

(1) $9=3^{x+1}$.

$3^2=3^{x+1} \Rightarrow 2=x+1 \Rightarrow x=1$.

(2) $\left(\frac{1}{2}\right)^x=8$.

$2^{-x}=2^3 \Rightarrow x=-3$.

■

3.6.4 Properties of exponential functions

Let a be a positive number such that $a \neq 1$. If u and v are real numbers, the following properties are true.

	Exponential with base a	Natural exponential function
Product Property	$a^{u+v}=a^u a^v$	$e^{u+v}=e^u e^v$.
Quotient Property	$a^{u-v}=\frac{a^u}{a^v}$	$e^{u-v}=\frac{e^u}{e^v}$.
Power Propety	$a^{uv}=(a^u)^v$	$e^{uv}=(e^u)^v$.

■ **Example 3.31** Given that $10^\alpha=3$, $10^\beta=4$. Find $10^{\alpha+\beta}$, $10^{\alpha-\beta}$, $10^{-2\alpha}$ and $10^{\frac{\beta}{5}}$.

$10^{\alpha+\beta}=10^\alpha 10^\beta=12$.

$10^{\alpha-\beta}=\frac{10^\alpha}{10^\beta}=0.75$.

$10^{-2\alpha}=(10^\alpha)^{-2}=3^{-2}=\frac{1}{9}$.

$10^{\frac{\beta}{5}}=(10^\beta)^{\frac{1}{5}}=\sqrt[5]{4}$.

■

3.7 Logarithms

In mathematics, the **logarithm** is the inverse operation to **exponentiation**, just as division is the inverse of multiplication and vice versa. That means the logarithm of a number is the exponent to which another fixed number, the **base**, must be raised to produce that number. In simple cases the logarithm counts factors in multiplication. For example, the base 10 logarithm of 1 000 is 3, as 10 to the power 3 is 1 000.

Definition 3.7.1—Logarithmic function. For $x>0$, $b>0$ and $b=0$, if $x=b^y$ then $y=\log_b x$. So the function given by

$$f(x)=\log_b x$$

is called the **logarithmic function with base** b.

Ⓡ "$\log_b x$" is pronounced as "the logarithm of x to base b" or "the base$-b$ logarithm of x" or (most commonly) "the log, base b, of x".

Ⓡ The defining identity is given by

$$b^{\log_b x}=x.$$

■ **Example 3.32** Write the following equations in a logarithm form with a suitable base.

(1) $3^4=81$.

$\Rightarrow \log_3 81=4$.

(2) $2^3=8$.

$\Rightarrow \log_2 8=3$.

(3) $4^2=16$.

$\Rightarrow \log_4 16=2$.

(4) $\frac{1}{32}=\frac{1}{2^5}$.

$\Rightarrow \log_2 \frac{1}{32}=-5$.

■

■ **Example 3.33** Use the definition of logarithmic function to evaluate each logarithm at the indicated value of x.

(1) $f(x)=\log_2 x$, $x=32$

$f(32)=\log_2 32=5$, because $2^5=32$.

(2) $f(x)=\log_3 x$, $x=1$.

$f(1)=\log_3 1=0$, because $3^0=1$.

(3) $f(x)=\log_4 x$, $x=2$.

$f(2)=\log_4 2=\frac{1}{2}$, because $4^{\frac{1}{2}}=\sqrt{4}=2$.

(4) $f(x)=\log_{10} x$, $x=\frac{1}{100}$.

$f(\frac{1}{100})=\log_{10}\frac{1}{100}=-2$, because $10^{-2}=\frac{1}{10^2}=\frac{1}{100}$.

■

3.7.1 Graphs of logarithmic functions

Characteristics of graph of $y=\log_a x$, $a>1$

The basic characteristics of logarithmic graphs $y=\log_a x$, $a>1$ are summarized as follows.

- Domain $(0,\infty)$.
- Range $(-\infty,\infty)$.
- x-intercept $(1,0)$.

- Increasing.
- One-to-one, therefore has an inverse function.
- The graph gets close to the y-axis when x tends to 0.
- Reflection of graph of $y=a^x$ about the line $y=x$.

To sketch the graph of $\log_a x$ you can use the fact that the graphs of inverse functions are reflections of each other in the line $y=x$. For example, sketch the graphs of $f(x)=2^x$ and $g(x)=\log_2 x$. For $f(x)=2^x$, we construct a table of values as below.

x	-2	-1	0	1	2	3
$f(x)=2^x$	$\frac{1}{4}$	$\frac{1}{2}$	1	2	4	8

By plotting these points and connecting them with a smooth curve, you obtain the graph as show in Figure 3. 27. Since $g(x)=\log_2 x$ is the inverse of the function $f(x)=2^x$, the graph of g is obtained by plotting the points $(f(x), x)$ and connecting them with a smooth curve. The graph of g is a reflection of the graph of f in the line $y=x$ as shown in Figure 3. 27 as well.

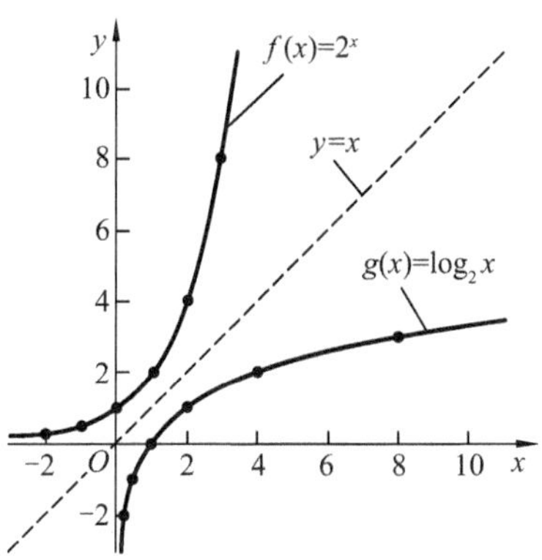

Figure 3. 27 The graphs of 2^x and $\log_2 x$.

3.7.2 Natural logarithmic functions

This inverse function of the **natural exponential function** $f(x)=e^x$ is called the **natural logarithmic function** and is denoted by the special symbol $\ln x$.

Ⓡ Note that the natural logarithm is written without a base. The base is understood to be the constant e.

As we know, the natural logarithmic function and the natural exponential function are inverse functions of each other. So, every logarithmic equation can be written in an equivalent exponential form, and every exponential equation can be written in logarithmic form. That is $f(x)=e^x$ and $g(x)=\ln x$ are inverse functions of each other, their graphs are reflections of each other in the line $y=x$ as illustrated in the Figure 3. 28 below.

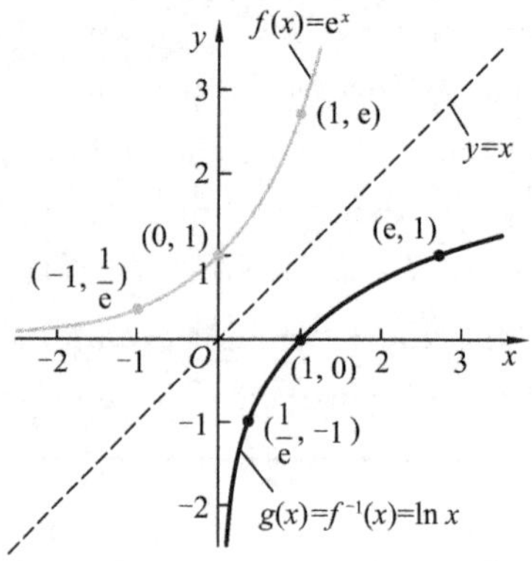

Figure 3.28 The graphs of e^x and $\ln x$.

3.7.3 Properties of logarithmic functions

Let a be a positive number such that $a\neq 1$, and let n be a real number. If u and v are positive real numbers, the following properties are true.

	Logarithm with base a	Natural logarithm
Product Property	$\log_a(uv)=\log_a u+\log_a v$	$\ln(uv)=\ln u+\ln v$.
Quotient Property	$\log_a\left(\frac{u}{v}\right)=\log_a u-\log_a v$	$\ln\left(\frac{u}{v}\right)=\ln u-\ln v$.
Power Propety	$\log_a u^n=n\log_a u$	$\ln u^n=n\ln u$.

Although common logarithms and natural logarithms are the most frequently used, you may occasionally need to evaluate logarithms with other bases. To do this, we may use the following **change-of-base formula**. Let a, b and x be positive real numbers such that $a\neq 1$ and $b\neq 1$ Then $\log_a x$ can be converted to a different base as follows:

(1) Base b.

$$\log_a x=\frac{\log_b x}{\log_b a}.$$

(2) Base 10.

$$\log_a x=\frac{\lg x}{\lg a}.$$

(3) Base e.

$$\log_a x=\frac{\ln x}{\ln a}.$$

Ⓡ In the right-hand side, one has only logarithms in the new base. In the left-hand side, x is above a and consequently in the right-hand side the logarithm of x is above the logarithm of a.

■ **Example 3.34** Changing bases using natural logarithms:

(1) $\log_4 25=\frac{\ln 25}{\ln 4}\approx 2.3219$.

(2) $\log_2 12=\dfrac{\ln 12}{\ln 2}\approx 3.5850$.

■

Exercise 3.8 Using change-of-base formula, prove that

(1) $\log_2 3\times\log_3 5\times\log_5 16=4$;

(2) $\log_{2\sqrt{2}} 3\sqrt{3}=\log_2 3$;

(3) $\log_7 3\times\log_9 5=\log_9 3\times\log_7 5$;

(4) $\log_4 9=\log_8 27$.

■

Proof

(1) We convert the logarithms in other bases to logarithms in base 2 and obtain

$$\log_2 3\times\log_3 5\times\log_5 16=\log_2 3\times\frac{\log_2 5}{\log_2 3}\times\frac{\log_2 16}{\log_2 5}=\log_2 16=4.$$

(2) We convert the logarithm in base $2\sqrt{2}$ to a logarithm in base 2 and obtain

$$\log_{2\sqrt{2}} 3\sqrt{3}=\frac{\log_2 3\sqrt{3}}{\log_2 2\sqrt{2}}=\frac{\log_2 3^{\frac{3}{2}}}{\log_2 2^{\frac{3}{2}}}=\frac{\frac{3}{2}\log_2 3}{\frac{3}{2}\log_2 2}=\frac{\log_2 3}{\log_2 2}=\log_2 3.$$

(3) We convert all logarithms to natural logarithms. We obtain

$$\log_7 3\times\log_9 5=\frac{\ln 3}{\ln 7}\times\frac{\ln 5}{\ln 9},$$

$$\log_9 3\times\log_7 5=\frac{\ln 3}{\ln 9}\times\frac{\ln 5}{\ln 7},$$

and the equality follows.

(4) We convert both logarithms to logarithms in base 2 (since both 4 and 8 are powers of 2). We obtain

$$\log_4 9=\frac{\log_2 9}{\log_2 4}=\frac{\log_2 3^2}{\log_2 2^2}=\frac{2\log_2 3}{2\log_2 2}=\frac{\log_2 3}{\log_2 2}=\log_2 3,$$

$$\log_8 27=\frac{\log_2 27}{\log_2 8}=\frac{\log_2 3^3}{\log_2 2^3}=\frac{3\log_2 3}{3\log_2 2}=\frac{\log_2 3}{\log_2 2}=\log_2 3,$$

and the equality follows.

■

Exercise 3.9 Expand each logarithmic expression:

(1) $\log_4(5x^3y)$;

(2) $\ln\dfrac{\sqrt{3x-5}}{7}$.

■

Solution

(1) $\log_4(5x^3y)=\log_4 5+\log_4 x^3+\log_4 y=\log_4 5+3\log_4 x+\log_4 y.$

(2) $\ln\dfrac{\sqrt{3x-5}}{7}=\ln\dfrac{(3x-5)^{1/2}}{7}=\ln(3x-5)^{1/2}-\ln 7=\dfrac{1}{2}\ln(3x-5)-\ln 7.$

■

Exercise 3.10 Condense each logarithmic expression:

(1) $\dfrac{1}{2}\lg x+3\lg(x+1)$;

(2) $2\ln(x+2)-\ln x$;

(3) $\dfrac{1}{3}[\log_3 x+\log_3(x+4)]$.

■

Solution

(1) $\dfrac{1}{2}\lg x+3\lg(x+1)=\lg x^{1/2}+\lg(x+1)^3=\lg[\sqrt{x}(x+1)^3].$

(2) $2\ln(x+2)-\ln x=\lg\left[\dfrac{(x+1)^2}{x}\right].$

(3) $\dfrac{1}{3}[\log_3 x+\log_3(x+4)]=\dfrac{1}{3}\log_3[x(x+4)]=\log_3[x(x+4)]^{\frac{1}{3}}=\log_3\sqrt[3]{x(x+4)}.$

■

3.8 Summary

Relations and mappings

A **relation** is a correspondence between two sets where each element in the first set, called the **domain**, corresponds to at least one element in the second set, called the **codomain.**

The Cartesian product of two sets X and Y is the set of all ordered pairs, written (x,y), where x is an element of X and y is an element of Y. Then x and y are called the **components** of the ordered pair.

The domain is the set of all the first components of the ordered pairs, and the **range** (which may not always be the same as the codomain) is the set of all the second components of the ordered pairs.

Using what is called **mapping notation**, we might simply write a relation between X and Y as $X \to Y$.

Functions

A function f from X to Y is a subset of the Cartesian product $X\times Y$ subject to the following condition: every element of X is the first component of one and only one ordered pair in the subset. In other words, for every x (called the **argument** or the **independent** variable) in X there is exactly one element y (called the **image** of x, or the **dependent variable**, since it depends on x) such that the ordered pair (x,y) is contained in the subset defining the function f.

The **domain** of a function is the complete set of all possible values of the independent variable. The **codomain** or **target set** of a function is the set into which all outputs of the function are constrained to fall. A function whose domain and codomain are both subsets of the set of real numbers is called a **numeric function.**

When a numeric function f is defined, it is often the case that the domain of f is not explicitly stated. In such a case, we shall consider the domain to be the set of real numbers x so that the value of $f(x)$ is also a real number. This is called the **implied domain** (or the **maximal domain**) of f. The **range** of a function is the complete set of all possible resulting values of the dependent variable (y, usually), after we have substituted the argument of the function with all possible values in the domain.

Characteristics of a function from a set *X* to a set *Y*

- Each element in X must be matched with an element in Y.
- Some elements in Y may not be matched with any element in X.
- Two or more elements in X may be matched with the same element in Y.
- An element in X (the domain) cannot be matched with two different elements in Y.

It is important to note that

- $f(x)$ does not mean f times x.
- The most common function names are f and F since the word function begins with "f". Other common function names are g and G, but any letter can be used.
- The letter most commonly used for the independent variable is x. The letter t is also common because it represents time in real-world applications, but any letter can be used.
- Although we can think of y and $f(x)$ as interchangeable, the function notation is useful when we want to consider two or more functions of the same independent variable or when we want to evaluate a function at more than one argument.

Four ways to represent a function

- *Verbally*, through a sentence that describes how the input variable is related to the output variable.
- *Numerically*, through a table or a list of ordered pairs that matches input values with output values.
- *Graphically*, through points on a graph in a coordinate plane in which the input values are represented by the horizontal axis and the output values are represented by the vertical axis.
- *Algebraically*, through a function described via some specific formula.

Injective, surjective and bijective functions

In mathematics, an **injective function** or **injection** or **one-to-one function** is a function that preserves distinctness: it never maps distinct elements of its domain to the same element of its codomain. In other words, every element of the function's range is the image of at most one element of its domain.

A function f from a set X to a set Y is **surjective** (or **onto**), or a **surjection**, if for every element y in the codomain Y of f there is at least one element x in the domain X of f such that $f(x)=y$. It is not required that x is unique; the function f may map one or more elements of X to the same element of Y.

A **bijection**, **bijective function** or **one-to-one correspondence** is a function between the elements of two sets, where each element of one set is paired with exactly one element of the other set, and each element of the other set is paired with exactly one element of the first set.

Hybrid functions

In mathematics, a **piecewise-defined function**, also called a **piecewise function** or a **hybrid function**, is a function which is defined by multiple sub-functions, each sub-function applying to a certain interval of the main function's domain.

Inverse functions

In mathematics, an inverse function is a function that "reverses" another function: if the function f applied to an input x gives a result of y, then applying its inverse function g to y gives the result x, and vice versa.

How to find the inverse of a function

To find the inverse of a function, express the function in a the form $y=f(x)$, interchange x and y, and then rearrange the equation to make y the subject of the equation again. If f is an invertible function with domain X and range Y, then its inverse f^{-1} has domain Y and range X.

How to graph the inverse of a function

The graphs of a function f and its inverse function f^{-1} are related to each other in the following way. If the point (a,b) lies on the graph of f then the point (b,a) must lie on the graph of f^{-1} and vice versa. This means that the graph of f^{-1} is a reflection of the graph of f in the line $y=x$.

Arithmetic combination of functions

Just as two real numbers can be combined using arithmetic operations, the output values of two or more functions can be combined using addition, subtraction, multiplication and division.

Composition of functions

In mathematics, composition of functions is the pointwise application of one function to the result of another to produce a third function.

The composition of the function f with function g is

$$(f\circ g)(x)=f(g(x)) .$$

The domain of $f\circ g$ is the set of all x in the domain of g such that $g(x)$ is in the domain of f.

The inverse of a composite function

The inverse of $f\circ g$ is $(f\circ g)^{-1}(x)$, obtained by reversing each of the individual component functions

$$g^{-1}\circ f^{-1}(x) \xleftarrow{g^{-1}} f^{-1}(x) \xleftarrow{f^{-1}} x.$$

So values of x are put into f^{-1}, and then values of $f^{-1}(x)$ are put into g^{-1}. This gives

$$(f\circ g)^{-1}(x)\equiv g^{-1}\circ f^{-1}(x).$$

Graph shifting

(1) Vertical shift c unit *upward* $h(x)=f(x)+c$.

(2) Vertical shift c unit *downward* $h(x)=f(x)-c$.

(3) Horizontal shift c unit to the *right* $h(x)=f(x-c)$.

(4) Horizontal shift c unit to the *left* $h(x)=f(x+c)$.

Graph reflections

The second common type of transformation is a **reflection.** Reflections in the coordinate axes of the graph of $y=f(x)$ are represented as follows.

(1) Reflection in the x-axis, $h(x)=-f(x)$.

(2) Reflection in the y-axis, $h(x)=f(-x)$.

Graph shrinking and stretching

Horizontal shifts, vertical shifts, and reflections are **rigid transformations** because the basic shape of the graph is unchanged. These transformations change only the position of the graph in the coordinate plane.

Non-rigid transformations are those that cause a *distortion*-a change in the shape of the original graph. For instance, a non-rigid transformation of the graph of $y=f(x)$ is represented by $g(x)=cf(x)$, where the transformation is a **vertical stretch** if $c>1$ and a **vertical shrink** if $0<c<1$. Another non-rigid transformation of the graph of $y=f(x)$ is represented by $h(x)= f(cx)$, where the transformation is a **horizontal shrink** if $c>1$ and a

horizontal stretch if $0<c<1$.

Even and odd functions

A function is said to be **even** if its graph is symmetric with respect to the y-axis and to be odd if its graph is symmetric with respect to the **origin.**

Test for even and odd functions

- A function $y=f(x)$ is **even** if, for each x in the domain of f, $f(-x)=f(x)$.
- A function $y=f(x)$ is **odd** if, for each x in the domain of f, $f(-x)=-f(x)$.

Periodic functions

A function f is said to be **periodic** with period P (P being a nonzero constant) if we have

$$f(x+P)=f(x)$$

for all values of x in the domain. If there exists a least positive constant P with the property that $f(x+P)=f(x)$ for all values of x in the domain, it is called the **fundamental period** (also **primitive period**, **basic period**, or **prime period**). If a function f is periodic with period P, then for all x in the domain of f and all positive integers n,

$$f(x+nP)=f(x).$$

The exponential function

The exponential function f with base a is denoted by

$$f(x)=a^x,$$

where $a>0, a\neq 1$ and x is any real number.

Characteristics of Graph of $y=a^x$, $a>1$

- Domain $(-\infty, \infty)$.
- Range $(0, \infty)$.
- y-intercept $(0,1)$.
- Increasing.
- The graph gets close to the x-axis for x tending to $-\infty$.

Characteristics of Graph of $y=a^{-x}$, $a>1$

- Domain $(-\infty, \infty)$.
- Range $(0, \infty)$.
- y-intercept $(0,1)$.
- Decreasing.
- The graph gets close to the x-axis for x tending to ∞.

The exponential function as a series

The number e can also be defined as

$$e = 1 + \frac{1}{1!} + \frac{1}{2!} + \frac{1}{3!} + \cdots = \sum_{k=0}^{\infty} \frac{1}{k!}.$$

In general, the function e^x which is often referred to as the **exponential function** can be defined as

$$e = 1 + \frac{x}{1!} + \frac{x^2}{2!} + \frac{x^3}{3!} + \cdots = \sum_{k=0}^{\infty} \frac{x^k}{k!}.$$

Properties of exponential functions

Let a be a positive number such that $a \neq 1$. If u and v are real numbers, the following properties are true.

	Exponential with Base a	Natural exponential function
Product Property	$a^{u+v} = a^u a^v$	$e^{u+v} = e^u e^v$.
Quotient Property	$a^{u-v} = \frac{a^u}{a^v}$	$e^{u-v} = \frac{e^u}{e^v}$.
Power Propety	$a^{uv} = (a^u)^v$	$e^{uv} = (e^u)^v$.

The logarithmic function

For $x>0, b>0$ and $b \neq 0$, if $x = b^y$ then $y = \log_b x$. So the function given by

$$f(x) = \log_b x$$

is called the **logarithmic function with base** b.

Characteristics of graph of $y = \log_a x$, $a>1$

The basic characteristics of logarithmic graphs $y = \log_a x$, $a>1$ are summarized as follows.

- Domain $(0, \infty)$.
- Range $(-\infty, \infty)$.
- x-intercept $(1, 0)$.
- Increasing.
- One-to-one, therefore has an inverse function.
- The graph gets close to the y-axis when x tends to 0.
- Reflection of graph of $y = a^x$ about the line $y = x$.

This inverse function of the **natural exponential function** $f(x) = e^x$ is called the **natural logarithmic function** and is denoted by the special symbol $\ln x$.

Properties of logarithmic functions

Let a be a positive number such that $a \neq 1$, and let n be a real number. If u and v are positive real numbers, the following properties are true.

	Logarithm with base a	Natural logarithm
Product Property	$\log_a(uv)=\log_a u+\log_a v$	$\ln(uv)=\ln u+\ln v$.
Quotient Property	$\log_a\left(\frac{u}{v}\right)=\log_a u-\log_a v$	$\ln\left(\frac{u}{v}\right)=\ln u-\ln v$.
Power Propety	$\log_a u^n=n\log_a u$	$\ln u^n=n\ln u$.

Change-of-base formula. Let a, b and x be positive real numbers such that $a\neq 1$ and $b\neq 1$. Then $\log_a x$ can be converted to a different base as follows:

(1) Base b.

$$\log_a x=\frac{\log_b x}{\log_b a}.$$

(2) Base 10.

$$\log_a x=\frac{\lg x}{\lg a}.$$

(3) Base e.

$$\log_a x=\frac{\ln x}{\ln a}.$$

3.9 Further problems

Problem 3.1 Determine whether each relation is a function.

(1) $\{(0,-3),(0,3),(-3,0),(3,0)\}$; (2) $x^2+y^2=9$;

(3) $y=x^2$.

Problem 3.2 Given the functions $f(x)=2x-3$ and $g(t)=5+t$, evaluate the quantities given below.

(1) $f(-2)$; (2) $g(1)$;

(3) $3f(-2)-2g(1)$; (4) $f(x+1)-f(x-1)$;

(5) $g(x+a)-f(x+a)$; (6) $\frac{f(x+h)-f(x)}{h}$;

(7) $\frac{g(t+h)-g(t)}{h}$.

Problem 3.3 Given the functions $f(x)=2x+3$ and $g(x)=-2x-1$, prove that $f(3)+g(3)=f(-3)+g(-3)$.

Problem 3.4 Given the function $f(x)=x-3$, evaluate the expression

$$E=f(1)\times f(2)\times\cdots\times f(99)\times f(100).$$

Problem 3.5 Given the function $f(x)=\frac{x}{x+1}$, evaluate the expression

$$E=f(1)\times f(2)\times\cdots\times f(99)\times f(100).$$

Problem 3.6 Given the functions $f(x)=ax+b$ and $g(x)=bx+a$, prove that if $f(-1)=g(-1)$ then the functions f and g coincide.

Problem 3.7 Find the domain of the given function and then express the domain in interval notation.

(1) $f(x)=2x-5$;
(2) $p(x)=\dfrac{x+2}{x-5}$;
(3) $T(x)=\dfrac{2}{x^2-4}$;
(4) $F(x)=\dfrac{1}{x^2+1}$;
(5) $q(t)=\sqrt{7-x}$;
(6) $G(t)=\sqrt{t^2-4}$;
(7) $F(x)=\dfrac{1}{\sqrt{x-3}}$;
(8) $f(x)=\sqrt[3]{1-2x}$.

Problem 3.8 For each of the following functions, find its inverse, and state the domain and range of f^{-1}.

(1) $f(x)=-3x+2$;
(2) $f(x)=x^3+1$;
(3) $f(x)=\sqrt{x-3}$;
(4) $f(x)=x^2-1$, $x\geqslant 0$;
(5) $f(x)=\dfrac{2}{x}$;
(6) $f(x)=\dfrac{2}{3-x}$;
(7) $f(x)=\dfrac{7x+1}{5-x}$.

Problem 3.9 Given the functions f and g, find $f+g$, $f-g$, $f\times g$, and $\dfrac{f}{g}$.

(1) $f(x)=2x+1$, $g(x)=1-x$;
(2) $f(x)=2x^2-x$, $g(x)=x^2-4$;
(3) $f(x)=\dfrac{1}{x}$, $g(x)=x$;
(4) $f(x)=\sqrt{x}$, $g(x)=2\sqrt{x}$;
(5) $f(x)=\sqrt{4-x}$, $g(x)=\sqrt{x+3}$.

Problem 3.10 For the given functions f and g, find the composite functions $f\circ g$ and $g\circ f$, and state their domains.

(1) $f(x)=2x+1$, $g(x)=x^2-3$;
(2) $f(x)=\dfrac{1}{x-1}$, $g(x)=x+2$;
(3) $f(x)=|x|$, $g(x)=\dfrac{1}{\sqrt{x-1}}$;
(4) $f(x)=\sqrt{x-1}$, $g(x)=x+5$;
(5) $f(x)=x^3+4$, $g(x)=(x-4)^{1/3}$.

Problem 3.11 Given the function $f(x)=\dfrac{1}{\sqrt[3]{x}}$, find $(f\circ f)(512)$.

Problem 3.12 Given the function $f(x)=\log_2 x$, find $(f\circ f)(256)$.

Problem 3.13 Find the function $f(x)=ax+b$ such that $(f\circ f)(x)=4x+3$.

Problem 3.14 Use the graph of $f(x)=x^2$ to write an equation for each function whose graph is shown in Figure 3.29.

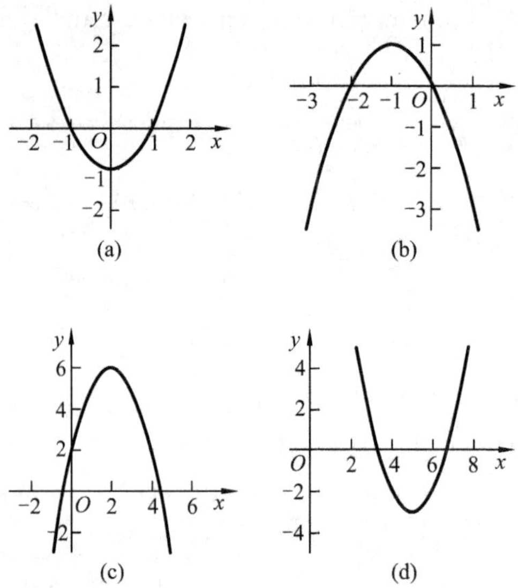

Figure 3.29 Use the graph of $f(x)=x^2$ to write the equations of the graphs above.

Problem 3.15 Determine whether the function is even, odd, or neither.

(1) $h(x)=x^2+2x$; (2) $f(x)=|x|+5$;

(3) $G(t)=|t-3|$; (4) $g(x)=\sqrt{x^2+x}$;

(5) $h(x)=\dfrac{1}{x}$.

Problem 3.16 Match the exponential function with its graphs labelled (a), (b), (c), and (d) in Figure 3.30.

(1) $f(x)=2^x$; (2) $f(x)=2^x+1$;

(3) $f(x)=2^{-x}$; (4) $f(x)=2^{x-2}$.

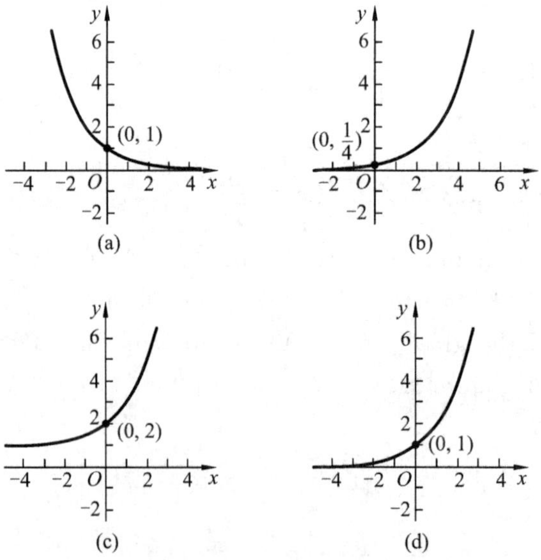

Figure 3.30 Four exponential functions.

Problem 3.17 Solve the following equation for x.

(1) $3^{(x+1)}=27$; (2) $\left(\frac{1}{2}\right)^{x}=32$;

(3) $e^{(3x+2)}=e^{3}$; (4) $e^{(x^{2}-3)}=e^{2x}$;

(5) $2^{(x-3)}=16$; (6) $5^{(x-2)}=\frac{1}{125}$;

(7) $e^{2x-1}=e^{4}$.

Problem 3.18 Rewrite the logarithm as a ratio of common logarithms and natural logarithms.

(1) $\log_{5}16$; (2) $\log_{3}47$;

(3) $\log_{\frac{1}{5}}x$; (4) $\log_{x}\frac{3}{10}$.

Problem 3.19 Use the properties of logarithms to expand the expression as a sum, difference, and /or constant multiple of logarithms.

(1) $\ln 4x$; (2) $\log_{8}x^{4}$;

(3) $\log_{5}\frac{5}{x}$; (4) $\ln\sqrt{z}$;

(5) $\ln xyz^{2}$; (6) $\ln z(z-1)^{2}$, $z>1$;

(7) $\log_{2}\frac{\sqrt{a-1}}{9}$, $a>1$; (8) $\ln\sqrt[3]{\frac{x}{y}}$;

(9) $\ln x^{2}\sqrt{\frac{y}{z}}$; (10) $\log_{5}\frac{x^{2}}{y^{2}z^{3}}$;

(11) $\ln\sqrt[4]{x^{3}(x^{2}+3)}$.

Problem 3.20 Condense the expression to the logarithm of a single quantity.

(1) $\ln 2+\ln x$; (2) $\log_{4}z-\log_{4}y$;

(3) $2\log_{2}x+4\log_{2}y$; (4) $\frac{2}{3}\log_{7}(z-2)$;

(5) $\frac{1}{4}\log_{3}5x$; (6) $\lg x-2\lg y+3\lg z$;

(7) $\ln x-[\ln(x+1)+\ln(x-1)]$; (8) $\frac{1}{3}[2\ln(x+3)+\ln x-\ln(x^{2}-1)]$;

(9) $\frac{1}{3}[\log_{8}y+2\log_{8}(y+4)]-\log_{8}(y-1)$.

Problem 3.21 Let $f:\{0,1,2,3\}\to\{2,3,4\}$ be defined by $f(0)=3$, $f(1)=2$, $f(2)=4$, $f(3)=2$. Prove that f is surjective, but it is not injective.

Problem 3.22 Let $f:\{0,1,2\}\to\{1,2,3,4\}$ be defined by $f(0)=1$, $f(1)=4$, $f(2)=3$. Prove that f is injective, but it is not surjective.

Problem 3.23 Let $f:\mathbf{Z}\to\mathbf{Z}$

$$f(x)=\begin{cases}\frac{x}{2}, & x \text{ even},\\ 2x, & x \text{ odd}.\end{cases}$$

Evaluate $f(0)$, $f(12)$, $f(3)$. Is f injective?

Problem 3.24 Prove that $f:\mathbf{Z}\to\mathbf{Z}$, $f(x)=x^2$ is neither injective nor surjective.

Problem 3.25 Solve the following equations:

(1) $5^{3x}=7^{x-2}$; (2) $2^{x+1}=3^x$.

Problem 3.26 Solve the following equations:

(1) $e^{2x}-5e^x+6=0$; (2) $e^{2x}-5e^x+4=0$.

Problem 3.27 Given the function $f(x)=\sqrt{x+1}$.

(1) Find the inverse function of f;

(2) Sketch the graphs of f and f^{-1} on the same set of coordinate axes;

(3) Describe the relationship between the graphs of f and f^{-1};

(4) State the domains and ranges of f and f^{-1}.

Problem 3.28 In the following functions:

(i) Identify the parent function f.

(ii) Describe the sequence of transformations from f to h.

(iii) Use function notation to write h in terms of f.

(1) $h(x)=x^2-9$. (2) $h(x)=-\sqrt{x}+4$.

(3) $h(x)=-(x+2)^2+3$. (4) $h(x)=-|-x+4|+6$.

— Answers —

Problem 3.1 (1) no. (2) no. (3) yes.

Problem 3.2 (1) -7. (2) 6. (3) -33. (4) 4. (5) $8-x-a$. (6) 2. (7) 1.

Problem 3.3 $f(3)+g(3)=f(-3)+g(-3)=2$.

Problem 3.4 Since $f(3)=0$, $E=0$.

Problem 3.5 By cancelling terms, $E=\frac{1}{101}$.

Problem 3.6 $f(-1)=g(-1)\Rightarrow -a+b=-b+a\Rightarrow a=b$. Then f coincides with g.

Problem 3.7 (1) $(-\infty,+\infty)$. (2) $(-\infty,5)\cup(5,+\infty)$.
(3) $(-\infty,-2)\cup(-2,2)\cup(2,+\infty)$. (4) $(-\infty,+\infty)$.
(5) $(-\infty,7]$. (6) $(-\infty,-2]\cup[2,+\infty)$.
(7) $(3,+\infty)$. (8) $(-\infty,+\infty)$.

Problem 3.8 The answers are listed in the following table.

Question	f^{-1}	Domain of f or Range of f^{-1}	Range of f or Domain of f^{-1}
(1)	$f^{-1}(x)=-\frac{x}{3}+\frac{2}{3}$	$(-\infty,\infty)$	$(-\infty,\infty)$
(2)	$f^{-1}(x)=\sqrt[3]{x-1}$	$(-\infty,\infty)$	$(-\infty,\infty)$
(3)	$f^{-1}(x)=x^2+3$	$[3,\infty)$	$[0,\infty)$
(4)	$f^{-1}(x)=\sqrt{x+1}$	$[0,\infty)$	$[-1,\infty)$
(5)	$f^{-1}(x)=\frac{2}{x}$	$(-\infty,0)\cup(0,\infty)$	$(-\infty,0)\cup(0,\infty)$
(6)	$f^{-1}(x)=3-\frac{2}{x}$	$(-\infty,3)\cup(3,\infty)$	$(-\infty,0)\cup(0,\infty)$
(7)	$f^{-1}(x)=\frac{5x-1}{x+7}$	$(-\infty,5)\cup(5,\infty)$	$(-\infty,-7)\cup(-7,\infty)$

Problem 3.9 (1) $f+g=x+2$, $f-g=3x$, $f\times g=-2x^2+x+1$, $\frac{f}{g}=\frac{2x+1}{1-x}$.

(2) $f+g=3x^2-x-4$, $f-g=x^2-x+4$, $f\times g=2x^4-x^3-8x^2+4x$, $\frac{f}{g}=\frac{2x^2-x}{x^2-4}$.

(3) $f+g=\frac{1+x^2}{x}$, $f-g=\frac{1-x^2}{x}$, $f\times g=1$, $\frac{f}{g}=\frac{1}{x^2}$.

(4) $f+g=3\sqrt{x}$, $f-g=-\sqrt{x}$, $f\times g=2x$, $\frac{f}{g}=\frac{1}{2}$.

(5) $f+g=\sqrt{4-x}+\sqrt{x+3}$, $f-g=\sqrt{4-x}-\sqrt{x+3}$, $f\times g=\sqrt{(4-x)(x+3)}$, $\frac{f}{g}=\frac{\sqrt{4-x}}{\sqrt{x+3}}$.

Problem 3.10 (1) $f\circ g(x)=2x^2-5$, Domain $(-\infty,\infty)$, $g\circ f(x)=4x^2+4x-2$, Domain $(-\infty,\infty)$.

(2) $f\circ g(x)=\dfrac{1}{x+1}$, Domain $(-\infty,-1)\cup(-1,\infty)$, $g\circ f(x)=\dfrac{1}{x-1}+2$, Domain $(-\infty,1)\cup(1,\infty)$.

(3) $f\circ g(x)=\dfrac{1}{\sqrt{x-1}}$, Domain $(1,\infty)$, $g\circ f(x)=\dfrac{1}{\sqrt{|x|-1}}$, Domain $(-\infty,-1)\cup(1,\infty)$.

(4) $f\circ g(x)=\sqrt{x+4}$, Domain $(-4,\infty)$, $g\circ f(x)=\sqrt{x-1}+5$, Domain $[1,\infty)$.

(5) $f\circ g(x)=x$, Domain $(-\infty,\infty)$, $g\circ f(x)=x$, Domain $(-\infty,\infty)$.

Problem 3.11 $(f\circ f)(x)=\sqrt[9]{x}\Rightarrow(f\circ f)(512)=2$.

Problem 3.12 $(f\circ f)(x)=\log_2(\log_2 x)\Rightarrow(f\circ f)(256)=3$.

Problem 3.13 $a=2$, $b=1$ or $a=-2, b=-3$.

Problem 3.14 (a) $y=x^2-1$. (b) $y=1-(x+1)^2$. (c) $y=-(x-2)^2+6$. (d) $y=(x-5)^2-3$.

Problem 3.15 (1) neither. (2) even. (3) neither. (4) neither. (5) odd.

Problem 3.16 (1) (d). (2) (c). (3) (a). (4) (b).

Problem 3.17 (1) $x=2$. (2) $x=-5$. (3) $x=1/3$. (4) $x=3$ or -1. (5) $x=7$. (6) $x=-1$. (7) $x=\dfrac{5}{2}$.

Problem 3.18 (1) $\dfrac{\lg 16}{\lg 5}, \dfrac{\ln 16}{\ln 5}$. (2) $\dfrac{\lg 47}{\lg 3}, \dfrac{\ln 47}{\ln 3}$.

(3) $\dfrac{\lg x}{\lg\frac{1}{5}}, \dfrac{\ln x}{\ln\frac{1}{5}}$. (4) $\dfrac{\lg\frac{3}{10}}{\lg x}, \dfrac{\ln\frac{3}{10}}{\ln x}$.

Problem 3.19 (1) $\ln 4+\ln x$. (2) $4\log_8 x$. (3) $1-\log_5 x$. (4) $\dfrac{1}{2}\ln z$. (5) $\ln x+\ln y+2\ln z$. (6) $\ln z+2\ln(z-1)$. (7) $\dfrac{1}{2}\log_2(a-1)-2\log_2 3$. (8) $\dfrac{1}{3}(\ln x-\ln y)$. (9) $2\ln x+\dfrac{1}{2}(\ln y-\ln z)$. (10) $2\log_5 x-(2\log_5 y+3\log_5 z)$. (11) $\dfrac{3}{4}\ln x+\dfrac{1}{4}\ln(x^2+3)$.

Problem 3.20 (1) $\ln 2x$. (2) $\log_4\dfrac{z}{y}$. (3) $\log_2(x^2y^4)$. (4) $\log_7\sqrt[3]{(z-2)^2}$. (5) $\log_3\sqrt[4]{5x}$. (6) $\lg\dfrac{xz^3}{y^2}$. (7) $\ln\left(\dfrac{x}{x^2-1}\right)$. (8) $\ln\sqrt[3]{\dfrac{x(x+3)^2}{x^2-1}}$. (9) $\log_8\dfrac{\sqrt[3]{y(y+4)^2}}{y-1}$.

Problem 3.21 f takes all the values in its codomain, so it is surjective. $f(1)=f(3)=2$, so f is not injective.

Problem 3.22 f does not repeat values, so it is injective. f does not take the value 2 from its codomain, so it is not surjective.

Problem 3.23 0, 6, 6, yes.

Problem 3.24 $f(-1)=f(1)=1$, so f is not injective. f does not take negative values, so f is not surjective.

Problem 3.25 (1) $-\dfrac{2\ln 7}{3\ln 5-\ln 7}$. (2) $\dfrac{\ln 2}{\ln 3-\ln 2}$.

Problem 3.26 (1) $x=\ln 2$, $\ln 3$. (2) $x=0$, $\ln 4$.

Problem 3.27 (1) $f^{-1}(x)=x^2-1$,

(2) The result is put in Figure 3.31.

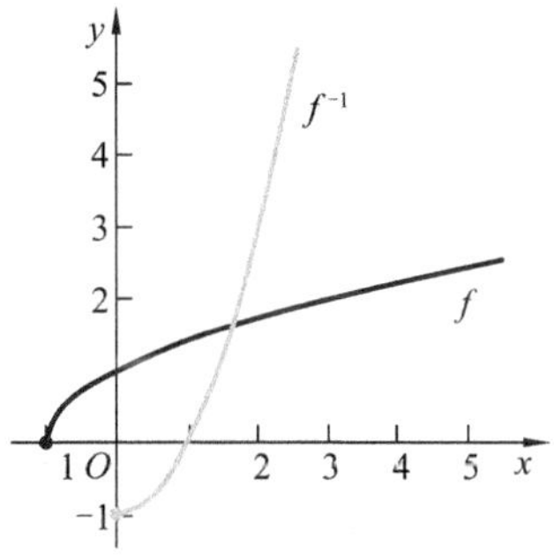

Figure 3.31 $f(x)=\sqrt{x+1}$ **and** $f^{-1}(x)=x^2-1$.

(3) The graph of f^{-1} is the reflection of the graph of f in the line $y=x$,

(4) Domain of f=Range of $f^{-1}=[-1,\infty)$, Range of f=Domain of $f^{-1}=[0,\infty)$.

Problem 3.28 (1) (i) $f(x)=x^2$. (ii) Vertical shift nine units downward. (iii) $h(x)=f(x)-9$.

(2) (i) $f(x)=\sqrt{x}$. (ii) Reflection in the x-axis and vertical shift four units upward. (iii) $h(x)=-f(x)+4$.

(3) (i) $f(x)=x^2$. (ii) Reflection in the x-axis, horizontal shift two units to the left, and vertical shift three units upward. (iii) $h(x)=-f(x+2)+3$.

(4) (i) $f(x)=|x|$. (ii) Reflections in the x-axis, horizontal shift four units to the right, and vertical shift six units upward. (iii) $h(x)=-f(-x+4)+6$.

Chapter 4 Algebra Ⅱ

4.1 Polynomial division

4.1.1 The concept of long division

Definition 4.1.1—Polynomial long division. In algebra, **polynomial long division** is an algorithm for dividing a polynomial by another polynomial of the same or lower degree.

In general terms, polynomial long division is an algorithm that implements the Euclidean division of polynomials, which starts from two polynomials $F(x)$ (the **dividend**) and $G(x)$ (the **divisor**) and produces, if $G(x)$ is not zero, a quotient $Q(x)$ and a remainder $R(x)$ such that

$$\frac{F(x)}{G(x)}=Q(x)+\frac{R(x)}{G(x)},$$

and either $R(x)=0$ or $\deg R<\deg G$ (the degree of $R(x)$ is lower than the degree of $G(x)$). These conditions uniquely define $Q(x)$ and $R(x)$, which means that $Q(x)$ and $R(x)$ do not depend on the method used to compute them. The result $R(x)=0$ occurs if and only if the polynomial $F(x)$ has $G(x)$ as a factor.

Ⓡ The above equality can also be written as

$$F(x)=Q(x)G(x)+R(x).$$

■ **Example 4.1** Consider the quadratic expression x^2+3x+2. It is known that

$$x^2+3x+2=(x+1)(x+2).$$

Let us now confirm the statement above using long division, with x^2+3x+2 as the dividend and $x+2$ as the divisor.

$$\begin{array}{r|l} & x+1 \\ \hline x+2 & x^2+3x+2 \\ & \underline{-x^2-2x} \\ & \quad\ \ x+2 \\ & \quad\ \underline{-x-2} \\ & \qquad\ \ 0 \end{array}$$

The intermediary steps are listed as follows.

Step 1 Divide the leading term of the dividend, x^2, by the leading term of the divisor, x, to obtain x. Multiply the result, x, by the divisor, $x+2$, to obtain x^2+2x.

Step 2 Subtract x^2+2x from x^2+3x to obtain x. Then bring down 2 and work with $x+2$ as the new dividend.

Step 3 Divide the leading term of the new dividend, x, by the leading term of the divisor, x to obtain 1. Multiply the result, 1, by the divisor, $x+2$ to obtain $x+2$. Now subtract $x+2$ from $x+2$ to get 0. The division stops here since the result is 0. This result, 0, is the remainder.

The quotient is then $x+1$ and the remainder is 0, yielding

$$\frac{x^2+3x+2}{x+2}=x+1+\frac{0}{x+2},$$

that is

$$x^2+3x+2=(x+1)(x+2).$$

■

Exercise 4.1 Find $\dfrac{2x^3+7x^2+7x+2}{2x+1}$.

■

Solution The dividend is $2x^3+7x^2+7x+2$ and the divisor is $2x+1$.

$$\begin{array}{r@{}l}
 & x^2+3x+2 \\
2x+1 \,\big) & 2x^3+7x^2+7x+2 \\
 & \underline{-2x^3-x^2} \\
 & \qquad 6x^2+7x \\
 & \qquad \underline{-6x^2-3x} \\
 & \qquad\qquad 4x+2 \\
 & \qquad\qquad \underline{-4x-2} \\
 & \qquad\qquad\qquad 0
\end{array}$$

The intermediary steps are listed as follows.

Step 1 Divide the leading term of the dividend, $2x^3$, by the leading term of the divisor, $2x$, to obtain x^2. Multiply the result, x^2, by the dividend, $2x+1$, to obtain $2x^3+x^2$.

Step 2 Subtract $2x^3+x^2$ from $2x^3+7x^2$ to obtain $6x^2$, bring down $+7x$ and work with $6x^2+7x$ as the new dividend.

Step 3 Divide the leading term of the new dividend, $6x^2$, by the leading term of the divisor, $2x$, to obtain $3x$. Multiply the result, $3x$, by $2x+1$ to obtain $6x^2+3x$.

Step 4 Subtract $6x^2+3x$ from $6x^2+7x$ to obtain $4x$, bring down 2 and work with $4x+2$ as the new dividend.

Step 5 Divide the leading term of the new dividend, $4x$, by the leading term of the divisor, $2x$, to obtain 2. Multiply 2 by $2x+1$ to obtain $4x+2$.

Step 6 Subtract $4x+2$ from $4x+2$ to obtain 0. The division stops here since the result is 0. 0 is the remainder.

The quotient is x^2+3x+2 and the remainder is zero, yielding

$$\frac{2x^3+7x^2+7x+2}{2x+1}=x^2+3x+2.$$

■

Ⓡ The division of polynomials is part of the process of determining their zeros. These are the points where the curve generated by the polynomial cuts the x-axis.

■ **Example 4.2** Let us determine the zeros of the polynomial

$$2x^3+7x^2+7x+2.$$

Let us first rewrite the polynomial as follows

$$2x^3+7x^2+7x+2=(2x+1)(x^2+3x+2).$$

To obtain the zeros of this polynomial, we must solve the equation

$$(2x+1)(x^2+3x+2)=0.$$

The polynomial x^2+3x+2 can be factorized further to give two linear factors, $x+1$ and $x+2$. We then obtain that

$$(2x+1)(x+1)(x+2)=0.$$

Solving for x, we obtain $x=\frac{1}{2}$ or $x=-1$ or $x=-2$.

■

■ **Example 4.3** Divide x^3+2x^2+3x+4 by $x+1$.

The dividend is x^3+2x^2+3x+4 and the divisor is $x+1$.

$$\begin{array}{r}
x^2+\ x+2 \\
x+1\overline{)\ x^3+2x^2+3x+4} \\
\underline{-x^3\ \ -x^2\qquad\quad} \\
x^2+3x\quad \\
\underline{-x^2\ \ -x\quad} \\
2x+4 \\
\underline{-2x-2} \\
2
\end{array}$$

Then,

$$\frac{x^3+2x^2+3x+4}{x+1}=x^2+x+2+\frac{2}{x+1}.$$

To verify the accuracy of the above result, multiply the quotient by the divisor and add the remainder to see that

$$[(x^2+x+2)(x+1)]+2=(x^3+2x^2+3x+2)+2=x^3+2x^2+3x+4.$$

■

■ **Example 4.4** Divide $5x^4+3x^3+x+1$ by x^2+1.

The dividend is $5x^4+3x^3+x+1$ and the divisor is x^2+1.

$$
\begin{array}{r}
5x^2+3x-5 \\
x^2+1\overline{)\ 5x^4+3x^3 \qquad\quad +x+1} \\
\underline{-5x^4 \qquad -5x^2} \qquad\quad \\
3x^3-5x^2+x \quad \\
\underline{-3x^3 \qquad -3x} \quad \\
-5x^2-2x+1 \\
\underline{5x^2 \qquad +5} \\
-2x+6
\end{array}
$$

The intermediary steps are listed as follows.

Step 1 Take note that the dividend is missing the term in x^2, so we can insert $0x^2$ in the dividend or leave a blank space in the corresponding place as in the solution above.

Step 2 Divide the leading term of the dividend, $5x^4$, by the leading term of the divisor, x^2, to obtain $5x^2$. Multiply $5x^2$ by x^2+1. The result is $5x^4+5x^2$. Also, there is no term in x^3, so we insert a term $0x^3$ or leave a blank space.

Step 3 Subtract coefficients of similar terms $(5x^4+3x^3+0x^2)-(5x^4+0x^3+5x^2)$ to obtain $3x^3-5x^2$, then bring down x and work with $3x^3-5x^2+x$ as the new dividend.

Step 4 Divide the leading term of the new dividend, $3x^3$, by the leading term of the divisor, x^2, to obtain $3x$. Multiply $3x$ by x^2+1 to obtain $3x^3+3x$.

Step 5 Subtract $3x^3+0x^2+3x$ from $3x^3-5x^2+x$ to obtain $-5x^2-2x$, then bring down 1 and work with $-5x^2-2x+1$ as the new dividend.

Step 6 Divide the leading term of the new dividend, $-5x^2$, by the leading term of the divisor, x^2, to obtain -5. Multiply -5 by x^2+1 and obtain $-5x^2-5$. The term in x is missing, so $0x$ should be inserted or a blank space should be left.

Step 7 Subtract $-5x^2+0x+1$ from $-5x^2-2x+1$ to obtain $-2x+6$. The division stops here since the result $-2x+6$ is of degree 1, smaller than the degree of the divisor, x^2+1, which is 2. $-2x+6$ is the remainder.

The result of the long division above can now be presented as

$$\frac{5x^4+3x^3+x+1}{x^2+1}=5x^2+3x-5+\frac{-2x+6}{x^2+1}.$$

■

4.1.2 The synthetic division

In algebra, **synthetic division** is a method of performing Euclidean division of polynomials, with less writing and fewer calculations than occur with polynomial long division. It is mostly taught for division by binomials of the form

$$x-a,$$

but the method generalizes to division by any polynomial with leading coefficient 1, and in fact to any polynomial. The advantages of synthetic division are that it allows one to calculate without writing variables, it uses few calculations, and it takes significantly less space on paper than long division. Also, the subtractions in long division are converted to additions by switching the signs at the very beginning, preventing sign errors.

■ **Example 4.5** Divide x^3+2x^2+3x+4 by $x+1$ using synthetic division.

$$\begin{array}{r|rrrr} -1 & 1 & 2 & 3 & 4 \\ & & -1 & -1 & -2 \\ \hline & 1 & 1 & 2 & 2 \end{array}$$

The intermediary steps (presented for division by polynomials of the form $x-a$) are listed as follows.

Step 1 Write the coefficients of the polynomial that is to be divided at the top.

Step 2 Negate the coefficients of the divisor.

Step 3 Write to the left of the vertical bar the last negated coefficient.

Step 4 Drop the first coefficient after the vertical bar to the last row.

Step 5 Multiply the dropped number by the number before the bar, and place it in the next column of the middle row.

Step 6 Perform an addition in that column and place the result in the last row.

Step 7 Repeat the previous two steps.

The remainder is the last number in the last row. The other numbers in the last row are coefficients of the quotient. To obtain the quotient, the terms are written with increasing degree, starting with the penultimate term in the last row, from right to left and with degree zero. The result of our division is that

$$\frac{x^3+2x^2+3x+4}{x+1}=x^2+x+2+\frac{2}{x+1}.$$

■

4.1.3 The polynomial remainder theorem

Definition 4.1.2—Polynomial remainder theorem. In algebra, the **polynomial remainder theorem** or **little Bézout's theorem** is an application of Euclidean division of polynomials. It states that if a polynomial $P(x)$ is divided by $x-\alpha$, then the remainder is the constant $P(\alpha)$.

Earlier on, we have seen that if F is the dividend, G is the divisor, Q is the quotient and R is the remainder, then $F(x)=Q(x)G(x)+R(x)$. In our specific case,

$$P(x)=(x-\alpha)Q(x)+R(x),\ R(x)=0 \text{ or } \deg(R)=0.$$

In both situations, $R(x)=R$ is a constant that is independent of x. Setting $x=\alpha$ in the above formula, we obtain that $P(\alpha)=R$.

■ **Example 4.6** Use the remainder theorem to find the remainder when

(1) x^3+2x^2+3x+4 is divided by $x+1$;

(2) $4x^2+3x+1$ is divided by $x-2$.

Let $P_1(x)=x^3+2x^2+3x+4$ and $P_2(x)=4x^2+3x+1$. We directly apply the remainder theorem to obtain that

(1) The remainder equals $P_1(-1)=(-1)^3+2(-1)^2-3+4=2$;

(2) The remainder equals $P_2(2)=4(2)^2+6+1=23$.

■

Exercise 4.2 When the polynomial $2x^3-bx^2+cx+2$ is divided by $x+1$ the remainder is -7, and when it is divided by $x-1$ the remainder is 5. Find the values of b and c.

■

Solution Let $F(x)=2x^3+bx^2+cx+2$. When $F(x)$ is divided by $x+1$, the remainder is $F(-1)=7$, and when it is divided by $x-1$ the remainder is $F(1)=5$. Solving these equations simultaneously, we obtain

$$\begin{cases} b-c=-7, \\ b+c=1. \end{cases}$$

■

Hence, $b=-3$ and $c=4$.

4.1.4 The factor theorem

Definition 4.1.3—Factor theorem. In algebra, the **factor theorem** is a theorem linking factors and zeros of a polynomial. It is a special case of the polynomial remainder theorem. The factor theorem states that $x-\alpha$ is a divisor of $P(x)$ if and only if $P(\alpha)=0$.

Two problems where the factor theorem is commonly applied are those of factoring a polynomial and finding the roots of a polynomial equation. It is a direct consequence of the theorem that these problems are essentially equivalent.

The factor theorem is also used to remove known zeros from a polynomial while leaving all unknown zeros intact, thus producing a lower degree polynomial whose zeros may be easier to find. Conceptually, the method works as follows.

(1) Find a zero α of the polynomial f.

(2) Use the factor theorem to conclude that $(x-\alpha)$ is a factor of $f(x)$.

(3) Compute the polynomial $g(x)=\dfrac{f(x)}{x-a}$.

(4) Conclude that any root $x\neq\alpha$ of $f(x)=0$ is a root of $g(x)=0$ (and $x=\alpha$ may still be a root of g as well). Since the polynomial degree of g is one less than that of f, it is easier to find the remaining zeros by studying g.

■ **Example 4.7** Factorize $x^3+6x^2+11x+6$.

Let $F(x)=x^3+6x^2+11x+6$. To find the linear factors of $F(x)$, if they exist, we use trial and error, testing first all integer divisors of the constant 6 and see which one is a root. The integer divisors of 6 are $\pm1, \pm2, \pm3$ and ±6. Obviously, $F(-1)=F(-2)=F(-3)=0$ and therefore $x+1$, $x+2$, $x+3$ are divisors of F, by the factor theorem.

It follows that $F(x)=Q(x)(x+1)(x+2)(x+3)$, and since F is of polynomial degree 3, one actually has $F(x)=c(x+1)(x+2)(x+3)$ with c being a constant. By equating the coefficients of x^3 in the L. H. S. and R. H. S., it is seen that $c=1$ and consequently

$$F(x)=(x+1)(x+2)(x+3).$$

Alternatively, once the factor $x+1$ has been determined, one can find by long division that

$$x^3+6x^2+11x+6=(x+1)(x^2+5x+6).$$

The algebraic quotient x^2+5x+6 can now be easily factorized, since it is a polynomial of degree 2 with roots -2 and -3. Hence,

$$x^3+6x^2+11x+6=(x+1)(x+2)(x+3).$$

■

■ **Example 4.8** Prove that 1 is a double root of the polynomial $F(x)=x^3-3x+2$.

Since $F(1)=1^3-3(1)+2=0$, it follows by the factor theorem that $x-1$ is a factor of $F(x)$. By long division, one may find that

$$x^3-3x+2=(x-1)(x^2+x-2).$$

The polynomial x^2+x-2 can be easily factorized, since it is a polynomial of degree 2 with roots 1 and -2. Hence

$$x^3-3x+2=(x-1)(x-1)(x-2)=(x-1)^2(x-2).$$

Consequently, 1 is a double root (since the power of $x-1$ is 2), while 2 is a simple root.

■

4.2 Algebraic fractions

Definition 4.2.1—Algebraic fraction. In algebra, an **algebraic fraction** is a fraction whose numerator and denominator are algebraic expressions.

Two examples of algebraic fractions are $\dfrac{x-3}{3x^3-2x^2+1}$ and $\dfrac{\sqrt{x+2}}{x^4-x^3+2x^2-5x-6}$.

Definition 4.2.2—Rational fraction. A **rational fraction** is an algebraic fraction whose numerator and denominator are both polynomials.

■ **Example 4.9** $\dfrac{x-3}{3x^3-2x^2+1}$ is a rational fraction, but $\dfrac{\sqrt{x+2}}{x^4-x^3+2x^2-5x-6}$ is not, because the numerator contains a square root function.

■

Ⓡ In a rational fraction, the coefficients of the polynomials don't have to be rational numbers. For instance, $\dfrac{\sqrt{2}x+\sqrt{5}}{\sqrt{7}x^2+2}$ is still a rational fraction.

Definition 4.2.3—Proper fraction. A rational fraction $\dfrac{f(x)}{g(x)}$ is called **proper** if the degree of the numerator is strictly smaller than the degree of the denominator, that is, $\deg(f)<\deg(g)$, and **improper** otherwise.

■ **Example 4.10** The fraction $\frac{x-3}{3x^3-2x^2+1}$ is proper, since the degree of the numerator is 1 and the degree of the denominator is 3, while the rational fractions $\frac{x^3-3}{x^2-x+4}$ and $\frac{x^3-1}{3x^3-2x^2+1}$ are improper (in the first case, the degree of the numerator is strictly larger than the degree of the denominator, while in the second case they are equal).

■

Any improper rational fraction can be expressed as the sum of a polynomial (possibly constant) and a proper rational fraction through the use of polynomial long division. For example,

$$\frac{x^3+x^2+1}{x^2-5x+6}=(x+6)+\frac{24x-35}{x^2-5x+6}.$$

Ⓡ The sum of two proper rational fractions is a proper rational fraction.

Definition 4.2.4—Irrational fraction. An irrational fraction is a fraction that contains the variable under a fractional exponent.

An example of an irrational fraction is

$$\frac{\sqrt{x}-a}{\sqrt[3]{x}-\sqrt{x}}.$$

Definition 4.2.5—Rationalization. The process of transforming an irrational fraction to a rational fraction is known as **rationalization.**

Every irrational fraction in which the radicals are monomials may be rationalized by finding the least common multiple of the indices of the roots and substituting the variable for another variable with the least common multiple as exponent. In the example given, the least common multiple is 6, hence we can substitute $x=y^6$ to obtain

$$\frac{y^3-a}{y^2-y^3}.$$

Algebraic fractions follow the same rules applicable to numerical fractions and are subject to the same operations of addition, subtraction, multiplication and division.

■ **Example 4.11**

$$\begin{aligned}\frac{1}{x+1}+\frac{2}{x+2}&=\frac{1(x+2)}{(x+1)(x+2)}+\frac{2(x+1)}{(x+1)(x+2)}\\&=\frac{x+2+2x+2}{(x+1)(x+2)}\\&=\frac{3x+4}{(x+1)(x+2)}.\end{aligned}$$

■

Exercise 4.3 Express $\frac{4}{x-1}-\frac{3x+1}{2x^2+5x+1}$ as a single fraction.

■

Solution

$$\frac{4}{x-1}-\frac{3x+1}{2x^2+5x+1}=\frac{4(2x^2+5x+1)-(3x+1)(x-1)}{(x-1)(2x^2+5x+1)}$$
$$=\frac{(8x^2+20x+4)-(3x^2-2x-1)}{(x-1)(2x^2+5x+1)}$$
$$=\frac{5x^2+22x+5}{(x-1)(2x^2+5x+1)}.$$

■

The processes and techniques of dividing and multiplying algebraic expressions are quite similar to those employed for numerical fractions. For multiplication, identify a common factor for the numerator and the denominator. If it does exist, then cancel it and continue this step until there is nothing to cancel. The next step is to multiply the numerators on their own and then the denominators on their own. For division, change the division sign to multiplication and then invert the second fraction.

■ **Example 4.12** Simplify $\frac{3(x+5)}{7}\times\frac{1}{6x}$.

$$\frac{3(x+5)}{7}\times\frac{1}{6x}=\frac{x+5}{7}\times\frac{1}{2x}$$
$$=\frac{x+5}{14x}.$$

■

■ **Example 4.13** Express $\frac{2(x+3)}{5}\times\frac{1}{x^2-9}$ as a single fraction.

$$\frac{2(x+3)}{5}\times\frac{1}{x^2-9}=\frac{2(x+3)}{5}\times\frac{1}{(x+3)(x-3)}$$
$$=\frac{2}{5(x-3)}.$$

■

Exercise 4.4 Express $\frac{3x-6}{8x}\div\frac{5x^2-5x-10}{6x}$ as a single fraction.

■

Solution

$$\frac{3x-6}{8x}\div\frac{5x^2-5x-10}{6x}=\frac{3(x-2)}{8x}\times\frac{6x}{5(x+1)(x-2)}$$
$$=\frac{9}{20(x+1)}.$$

■

Partial fractions

Definition 4.2.6—Partial fractions. The process of expressing a proper rational fraction as the sum of two or more simpler fractions, whose denominators are powers either of monomials or of polynomials of degree 2 which cannot be further factored as products of monomials with real coefficients, is called resolving it into **partial fractions.**

■ **Example 4.14**

$$\frac{2x}{x^2-1}=\frac{1}{x-1}+\frac{1}{x+1}.$$

■

The technique of decomposing a fraction depends more on the nature of the denominator. In this section, we shall consider the following categories of denominators:

(1) Linear factors in the denominator.

(2) Quadratic factors in the denominator.

(3) Repeated factors in the denominator.

Start with a proper rational function $f(x)=\frac{P(x)}{Q(x)}$. By the fundamental theorem of algebra, one can factor the denominator as much as possible, group similar terms and obtain that

$Q(x)=(x-\alpha_1)^{j_1}(x-\alpha_2)^{j_2}\cdots(x-\alpha_m)^{j_m}(x^2+\beta_1x+\gamma_1)^{k_1}(x^2+\beta_2x+\gamma_2)^{k_2}\cdots(x^2+\beta_nx+\gamma_n)^{k_n}$, in which $\alpha_1,\cdots,\alpha_m,\beta_1,\cdots,\beta_n,\gamma_1,\cdots,\gamma_n$ are distinct constants with $\beta_i^2-4\gamma_i<0$ (that is, all $x^2+\beta_ix+\gamma_i$ can't be factored anymore), and $j_1,\cdots,j_m,k_1,\cdots,k_n$ are positive integers.

The partial fractions are then obtained by supposing that

$$f(x)=\frac{P(x)}{Q(x)}=\sum_{i=1}^{m}\sum_{r=1}^{j_i}\frac{A_{ir}}{(x-\alpha_i)^r}+\sum_{i=1}^{n}\sum_{r=1}^{k_i}\frac{B_{ir}x+C_{ir}}{(x^2+b_ix+c_i)^r}$$

(note that not only the maximal powers $(x-\alpha_i)^{j_i}$ and $(x^2+\beta_ix+\gamma_i)^{k_i}$ appear in the R.H.S., but all the intermediary powers as well; note also that the degrees of the numerators are 1 less than those of the "central units" (the content of the parenthesis) of the denominators), clearing denominators, expanding and equating the coefficients of terms involving the powers of x. A system with constants A_{ir}, B_{ir} and C_{ir} as unknowns is then obtained, which should then be solved and the numeric values of A_{ir}, B_{ir} and C_{ir} should be substituted back into the expression of f.

If one starts with $f(x)=\frac{P(x)}{Q(x)}$ being an improper rational function, that is, $\deg(P)\geqslant\deg(Q)$, then it is necessary to perform the Euclidean division of P by Q, using polynomial long division. Then$=\frac{P(x)}{Q(x)}=E(x)+\frac{R(x)}{Q(x)}$ and then the above procedure is used for the fraction $\frac{R(x)}{Q(x)}$.

■ **Example 4.15** Express $\frac{x+1}{x^2-16}$ using partial fractions.

Rewrite $\frac{x+1}{x^2-16}$ with factorized denominator as $\frac{x+1}{(x+4)(x-4)}$. To be resolved into partial fractions, the above algebraic expression must be written as a sum of fractions which involve both factors of the denominator, appearing separately:

$$\frac{x+1}{(x+4)(x-4)}=\frac{A}{x+4}+\frac{B}{x-4}.$$

Clearing denominators shows that $x+1=A(x-4)+B(x+4)$. Expanding and equating the coefficients of powers of x gives

$$1=A+B \text{ and } 1=4(B-A).$$

Solving for A and B yields $A=\frac{3}{8}$ and $B=\frac{5}{8}$.

■

■ **Example 4.16** Express $\frac{x+2}{(x+1)(x-2)^2}$ using partial fractions.

To be resolved into partial fractions, the above fraction (involving the repeated factor $(x-2)^2$) must be written as a sum of fractions which involve the factors of the denominator, involving not only the maximal power $(x-2)^2$ but also the intermediary power $(x-2)$ as well:

$$\frac{x+2}{(x+1)(x-2)^2}=\frac{A}{x+1}+\frac{B}{x-2}+\frac{C}{(x-2)^2}.$$

Clearing denominators shows that $x+2=(A+B)x^2+(C-4A-B)x+4A-2B+C$. Equating the coefficients of powers of x gives

$$0=A+B,\ 1=C-4A-B \text{ and } 2=4A-2B+C.$$

Solving for A, B and C yields $A=\frac{1}{9}$, $B=-\frac{1}{9}$ and $C=\frac{4}{3}$.

■

■ **Example 4.17** Express $\frac{3x^2+3x+1}{(x+1)(x^2+1)}$ using partial fractions.

To be resolved into partial fractions, the above fraction must be written as a sum of fractions which involve both factors of the denominator, appearing separately (note that since one of the denominators, x^2+1, is a polynomial of degree 2, its corresponding numerator is of degree 1):

$$\frac{3x^2+3x+1}{(x+1)(x^2+1)}=\frac{A}{x+1}+\frac{Bx+C}{x^2+1}.$$

Clearing denominators shows that $3x^2+3x+1=A(x^2+1)+(Bx+C)(x+1)$. Expanding and equating the coefficients of powers of x gives

$$3=A+B,\ 3=B+C \text{ and } 1=A+C.$$

Solving for A, B and C yields $A=\frac{1}{2}$, $B=\frac{5}{2}$ and $C=\frac{1}{2}$.

■

■ **Example 4.18** Express the improper fraction $\frac{4x^3+10}{(x-1)(x+2)}$ using partial fractions.

Notice that the degree of the numerator is higher than the degree of the denominator. We first divide $4x^3+10$ by x^2+x-2 and then get

$$\frac{4x^3+10}{(x-1)(x+2)}=(4x-4)+\frac{2(6x+1)}{x^2+x-2}.$$

Now, the proper fraction in the R.H.S. can be resolved into partial fractions.

$$\frac{2(6x+1)}{x^2+x-2}=\frac{A}{x-1}+\frac{B}{x+2}.$$

Clearing denominators shows that $2(6x+1)=A(x+2)+B(x-1)$. Expanding and equating the coefficients of powers of x gives

$$12=A+B \text{ and } 2=2A-B.$$

Solving for A and B yields $A=\frac{14}{3}$ and $B=\frac{22}{3}$.

■

4.3 Curve sketching

In previous courses, you have learned to draw or sketch graphs by plotting a few points generated using a table of values. In this section we depart from that method. Curve sketching simply means here producing an outline graph that shows the general behaviour of a function or polynomial. The requirement is to show only the main features of the graph.

Definition 4.3.1—Asymptote. In analytic geometry, an asymptote of a curve is a line such that the distance between the curve and the line approaches zero as they tend to infinity.

The following steps should be followed when sketching a graph:

(1) Determine the intercepts of both the x-axis and the y-axis (shortened from now on as the x-intercept and the y-intercept), if any. These are the points where the graph crosses the x and y-axes, respectively.

(2) Determine the symmetries of the graph, if any.

(3) Determine any bounds on the values of x and y.

(4) Identify the asymptotes of the graph.

(5) Identify any readily available additional information, such as other points the graph passes through and tangents to the graph in these points.

■ **Example 4.19** Sketch the curve $y=x^2+3x+2$.

We start by identifying the main features of the graph as listed above (see Figure 4.1).

(1) The y-coordinate of the y-intercept is found by setting $x=0$ and calculating $y=(0)^2+3(0)+2=2$. The x-coordinates of the x-intercepts are found by setting $y=0$ and solving the equation $x^2+3x+2=0$, which leads to $x=-1$ and $x=-2$. Therefore, the graph passes through the points $(0,2)$, $(-1,0)$ and $(-2,0)$.

(2) Since $y=x^2+3x+2=(x+\frac{3}{2})^2-\frac{1}{4}$, one notes that the graph is symmetric about

$x=-\dfrac{3}{2}$.

(3) Since a square is always positive, the function defined over the set of real numbers has a global minimum point at $x=-\dfrac{3}{2}$, equal to $-\dfrac{1}{4}$. Increasing or decreasing the value of x, the value of y goes to ∞. There are no bounds on the values of x since the function is defined over the set of real numbers.

(4) As observed, the curve passes through $(0,2)$. The equation of the tangent to the graph in this point, obtained by removing all but the terms of lowest order from the equation, is $y=3x+2$. Note that this procedure is not generally valid, but furnishes correct results whenever the R.H.S. (x^2+3x+2 here) is a polynomial and the abscissa of the point is 0.

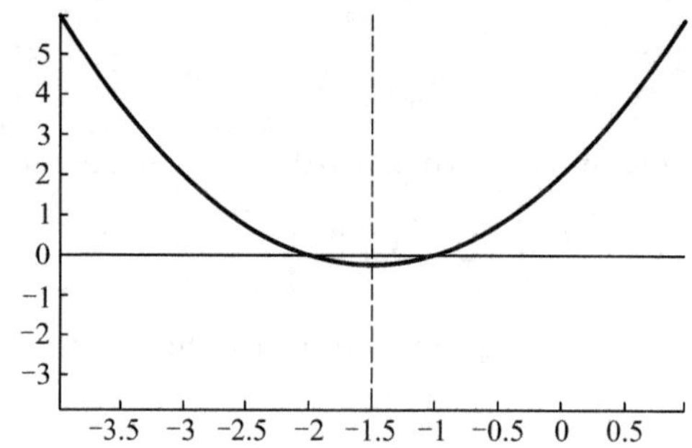

Figure 4.1 Graph of the function $y=x^2+3x+2$.

■

■ **Example 4.20** Sketch the curve $y=x^3+2x^2-5x-6$. We start by identifying the main features of the graph as listed above.

(1) The y-coordinate of the y-intercept is found by calculating $y=(0)^3+2(0)^2-5(0)-6=-6$. The x-coordinates of the x-intercepts are found by calculating $x^3+2x^2-5x-6=0$, that is, $x=-3$, $x=-4$ and $x=2$. Therefore, the graph passes through the points $(0,-6)$, $(-3,0)$, $(-4,0)$ and $(2,0)$.

(2) Since $y=x^3+2x^2-5x-6=(x-2)(x+3)(x+4)$, one notes that the graph is increasing for $x\in\left(\dfrac{-2+\sqrt{19}}{3},+\infty\right)\cup\left(-\infty,\dfrac{-2-\sqrt{19}}{3}\right)$, while decreasing for $x\in\left(\dfrac{-2-\sqrt{19}}{3},\dfrac{-2+\sqrt{19}}{3}\right)$ as shown in Figure 4.2.

(3) As shown in Figure 4.2, the function defined over the real number set has a local minimum point at $x=\dfrac{-2+\sqrt{19}}{3}$ and a local maximum point at $x=\dfrac{-2-\sqrt{19}}{3}$. When the value of x goes to ∞ , the value of y goes to ∞.

(4) The curve passes through $(0,-6)$. The equation of the tangent to the graph in this point is $y=-5x-6$.

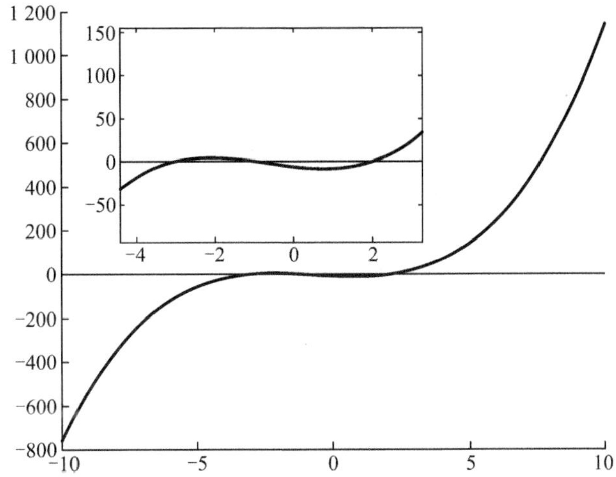

Figure 4.2 Graph of the function $y=x^3+2x^2-5x-6$.

■

■ **Example 4.21** Sketch the graph of $y=\frac{4}{x-3}$.

We start by identifying the main features of the graph as listed above.

(1) The y-coordinate of the y-axis intercept is found by setting $x=0$, which leads to $y=-3$, so the graph crosses the y-axis at $(0,-3)$. When $y=0$ then $0=\frac{4}{x-3}$ and this shows that there is no value of x for which y is zero, that is, there are no x-axis intercepts.

(2) To determine the vertical asymptote we find the value that makes the denominator zero. Then $x-3=0$, therefore $x=3$ as shown in Figure 4.3. Also, the graph has a horizontal asymptote $y=0$ because when x goes to $\pm\infty$, y approaches 0.

(3) The function defined over the set $(-\infty,3)\cup(3,+\infty)$ has two branches. The graph of each branch is decreasing.

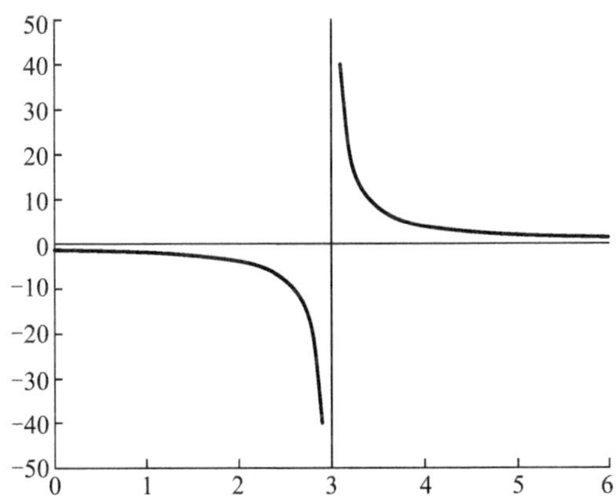

Figure 4.3 Graph of the function $y=\frac{4}{x-3}$.

■

4.4 Summary

The polynomial division

If we have two polynomials: a dividend $F(x)$ and a divisor $G(x)$ (where $G(x)\neq 0$), we can express the result of their division as follows:

$$\frac{F(x)}{G(x)}=Q(x)+\frac{R(x)}{G(x)},$$

where $Q(x)$ is the quotient and $R(x)$ is the remainder. If $R(x)=0$, then, $Q(x)$ and $G(x)$ are factors of $F(x)$.

The remainder theorem

If a polynomial $F(x)$ is divided by a linear factor $x-\alpha$ then the remainder is $F(\beta)$.

The factor theorem

If $F(\beta)=0$, then $F(x)$ divided by $x-\alpha$ does not leave a remainder, that is, $x-\beta$ is a linear factor of $F(x)$.

Partial fractions decomposition

The technique of decomposing a fraction depends more on the nature of the denominator. Start with a proper rational fraction $f(x)=\frac{P(x)}{Q(x)}$. By the fundamental theorem of algebra, one can obtain the decomposition

$Q(x)=(x-\alpha_1)^{j_1}(x-\alpha_2)^{j_2}\cdots(x-\alpha_m)^{j_m}(x^2+\beta_1x+\gamma_1)^{k_1}(x^2+\beta_2x+\gamma_2)^{k_2}\cdots(x^2+\beta_nx+\gamma_n)^{k_n}$,

in which $\alpha_1,\cdots,a_m,\beta_1,\cdots,\beta_n,\gamma_1,\cdots,\gamma_n$ are distinct constants with $\beta_i^2-4\gamma_i<0$. Then the partial fractions are generally obtained by supposing that

$$f(x)=\frac{P(x)}{Q(x)}=\sum_{i=1}^{m}\sum_{r=1}^{j_i}\frac{A_{ir}}{(x-a_i)^r}+\sum_{i=1}^{n}\sum_{r=1}^{k_i}\frac{B_{ir}x+C_{ir}}{(x^2+b_ix+c_i)^r}$$

clearing denominators, expanding and equating the coefficients of terms involving the powers of x. A system with constants A_{ir}, B_{ir} and C_{ir} as unknowns is then obtained, which should then be solved and the numeric values of A_{ir}, B_{ir} and C_{ir} should be substituted back into the expression of f.

If one starts with $f(x)=\frac{P(x)}{Q(x)}$being an improper rational function, that is, $\deg(P)\geqslant\deg(Q)$, then it is necessary to perform the Euclidean division of P by Q, using polynomial long division. Then $\frac{P(x)}{Q(x)}=E(x)+\frac{R(x)}{Q(x)}$ and then the above procedure is used for the fraction $\frac{R(x)}{Q(x)}$.

Curve sketching

The following are usually easy to carry out and give important clues as to the shape of a curve:

(1) Determine the x-intercepts and y-intercepts, if any.

(2) Determine the symmetries, if any.

(3) Determine any bounds on the values of x and y.

(4) Identify the asymptotes of the graph.

(5) Identify any readily available additional information, such as other points the graph passes through and tangents to the graph in these points.

4.5 Further problems

Problem 4.1 Using long division, divide

(1) x^2+x+1 by $x-1$;　(2) $2x^2+5x-3$ by $x-3$;

(3) $12x^3+2+11x+20x^2$ by $2x+1$;　(4) x^3+2x^2-3x+4 by $x-7$.

Problem 4.2 Using synthetic division, divide

(1) $2x^3+x^2+x+1$ by $x-1$;　(2) $7x^3+2x^2+x-1$ by $x+2$;

(3) x^3+x^2+x+1 by $x+9$;　(4) $3x^4-5x^2+3$ by $x+2$.

Problem 4.3 By combining the factor theorem and another appropriate result, factorize the following polynomials completely

(1) $x^3+8x^2+17x+10$;　(2) $2x^3+13x^2+23x+10$.

Problem 4.4 Find the zeros of the polynomial

(1) $3x^3+10x^2+9x+2$;　(2) x^3+5x^2+2x-8.

Problem 4.5 The polynomial $2x^3+bx^2+cx+2$ has factors $x+1$, $x+3$ and a third factor. Find the constants b and c.

Problem 4.6 Consider $F(x)=2x^3-7x^2-5x+4$.

(1) Find the remainder when $F(x)$ is divided by $(x-1)$.

(2) Use the factor theorem to show that $(x+1)$ is a factor of $F(x)$.

Problem 4.7 Consider $G(x)=x^3+ax^2+bx+3$, where a and b are constants. It is known that when $G(x)$ is divided by $(x+2)$ the remainder is 7.

(1) Show that $2a-b=6$.

(2) Find a and b knowing that, in addition to the above, when $G(x)$ is divided by $(x-1)$ the remainder is 4.

Problem 4.8 Simplify

(1) $\dfrac{3x}{4}\times\dfrac{2}{x}$;　(2) $\dfrac{x^2-1}{x}\times\dfrac{2x}{x+1}$;

(3) $\dfrac{6(x+4)}{x^2-4}\times\dfrac{x+2}{3(x^2-16)}$;　(4) $\dfrac{2x^2-4x}{x^2-2x+1}\times\dfrac{(x-1)^2}{2x}$.

Problem 4.9 Compute

(1) $\dfrac{1}{x}+\dfrac{2}{x+1}$;　(2) $\dfrac{4}{x+2}+\dfrac{3}{x+3}$;

(3) $\frac{7}{x-1}-\frac{2}{x-2}$; (4) $\frac{5}{x-2}-\frac{6}{x+3}$.

Problem 4.10 Compute and simplify

(1) $1+\frac{x^2+x-6}{(x+4)(x-2)}$; (2) $\frac{2}{2x-1}-\frac{1}{x+1}$;

(3) $\frac{x^2-8x+15}{2x^2-7x-15}$; (4) $\frac{6}{x-2}-\frac{6}{x+3}$.

Problem 4.11 Express the following using partial fractions:

(1) $\frac{5x+7}{(x+3)(x-1)}$; (2) $\frac{3}{x^2-x-2}$;

(3) $\frac{5x+10}{x^2+2x-3}$; (4) $\frac{7x-12}{x^2-3x+2}$.

Problem 4.12 Express the following using partial fractions:

(1) $\frac{3x^2+5x+6}{(x+1)(x^2+3)}$; (2) $\frac{4x^2+5x+9}{(x+1)(x^2+x+4)}$;

(3) $\frac{5x^2-2x+15}{x(x^2+3)}$; (4) $\frac{3x^2+20x+3}{(x-2)(x^2+3x+1)}$.

Problem 4.13 Express the following using partial fractions:

(1) $\frac{x+1}{(x+1)(x-1)^2}$; (2) $\frac{3x+5}{(x-1)^2}$;

(3) $\frac{5x+15}{(x-1)(x-4)^2}$; (4) $\frac{x+1}{(x+1)(x-3)^2}$.

Problem 4.14 Express the following using partial fractions:

(1) $\frac{3x^3+12}{(x-1)(x^2+1)}$; (2) $\frac{5x^3+15}{(x-2)(x+3)}$;

(3) $\frac{2x^3+4x^2+5x-5}{(x+1)^2}$; (4) $\frac{3x^3-2x+7}{(x-2)(x^2+1)}$.

Problem 4.15 Describe and sketch the graphs of the following functions:

(1) $f(x)=x^2-1$; (2) $f(x)=2x^2+5x+2$;

(3) $f(x)=3x^3+11x^2+12x+4$; (4) $f(x)=2x^3-9x^2+10x-3$.

Problem 4.16 Let us consider $H(x)=(2x-1)(x+2)(x-3)$.

(1) Express $H(x)$ in the form Ax^3+Bx^2+Cx+D, giving the values of the constants A, B, C and D.

(2) Find the value of the constant a, given that $(x+3)$ is a factor of $H(x)+ax$.

(3) Express $\frac{x-3}{H(x)}$ using partial fractions.

(4) Solve the inequality $H(x)>3x(x+2)(x-3)$.

Problem 4.17 A curve has the equation $y=\frac{2x+4}{(x-1)(2x+1)}$.

(1) Express $\frac{2x+4}{(x-1)(2x+1)}$ in partial fractions.

(2) Determine the stationary points of this curve.

Problem 4. 18 The cubic polynomial $f(x)=2x^3+bx^2-cx-6$ has factors $(x+1)$, $(x+2)$ and a third factor.

(1) Find the values of b and c.

(2) Factorize $f(x)$ completely.

(3) Sketch the graph of $f(x)$.

Problem 4. 19 The functions $f(x)$ and $g(x)$ are defined by

$$f(x):\ x\to\frac{1}{x},\ x\in\mathbf{R},\ x\neq 0,$$

$$g(x):\ x\to\frac{x-1}{x+1},\ x\neq -1,$$

respectively.

(1) Find $f(g(x))$.

(2) Find $f(g(x))^{-1}$.

(3) Explain the relationship between the graphs of $f(g(x))$ and $f(g(x))^{-1}$.

(4) Sketch the graph of $f(x)$.

—— Answers ——

Problem 4.1

(1)
$$
\begin{array}{r}
x+2 \\
x-1\overline{)\,x^2+x+1} \\
\underline{-x^2+x} \\
2x+1 \\
\underline{-2x+2} \\
3
\end{array}
$$

(2)
$$
\begin{array}{r}
2x\ +11 \\
x-3\overline{)\,2x^2+5x\ -3} \\
\underline{-2x^2+6x} \\
11x\ -3 \\
\underline{-11x+33} \\
30
\end{array}
$$

(3)
$$
\begin{array}{r}
6x^2\ +7x+2 \\
2x+1\overline{)\,12x^3+20x^2+11x+2} \\
\underline{-12x^3\ -6x^2} \\
14x^2+11x \\
\underline{-14x^2\ -7x} \\
4x+2 \\
\underline{-4x-2} \\
0
\end{array}
$$

(4)
$$
\begin{array}{r}
x^2\ +9x+60 \\
x-7\overline{)\,x^3+2x^2\ -3x\ +4} \\
\underline{-x^3+7x^2} \\
9x^2\ -3x \\
\underline{-9x^2+63x} \\
60x\ +4 \\
\underline{-60x+420} \\
424
\end{array}
$$

Problem 4.2

(1)
$$
\begin{array}{c|cccc}
1 & 2 & 1 & 1 & 1 \\
 & & 2 & 3 & 4 \\
\hline
 & 2 & 3 & 4 & 5
\end{array}
$$

(2)
$$
\begin{array}{c|cccc}
-2 & 7 & 2 & 1 & -1 \\
 & & -14 & 24 & -50 \\
\hline
 & 7 & -12 & 25 & -51
\end{array}
$$

(3)
$$
\begin{array}{c|cccc}
-9 & 1 & 1 & 1 & 1 \\
 & & -9 & 72 & -657 \\
\hline
 & 1 & -8 & 73 & -656
\end{array}
$$

(4)
$$
\begin{array}{c|ccccc}
-2 & 3 & 0 & -5 & 0 & 3 \\
 & & -6 & 12 & -14 & 28 \\
\hline
 & 3 & -6 & 7 & -14 & 31
\end{array}
$$

Problem 4.3

(1) $(x+1)(x+2)(x+5)$.

(2) $(x+2)(2x^2+9x+5)$.

Problem 4.4

(1) $-1, -2, -\frac{1}{3}$.

(2) $-4, -2, 1$.

Problem 4.5 $b=c=\frac{26}{3}$.

Problem 4.6

(1)
$$
\begin{array}{r}
2x^2\ -5x-10 \\
x-1\overline{)\,2x^3-7x^2\ -5x\ +4} \\
\underline{-2x^3+2x^2} \\
-5x^2\ -5x \\
\underline{5x^2\ -5x} \\
-10x\ +4 \\
\underline{10x-10} \\
-6
\end{array}
$$

(2)
$$
\begin{array}{r}
2x^2-9x+4 \\
x+1\overline{)\,2x^3-7x^2-5x+4} \\
\underline{-2x^3-2x^2} \\
-9x^2-5x \\
\underline{9x^2+9x} \\
4x+4 \\
\underline{-4x-4} \\
0
\end{array}
$$

Problem 4.7 (1) Since $G(-2)=7$, we obtain that $-8+4a-2b+3=7$, which implies that $2a-b=6$.

(2) Since $G(1)=4$, we obtain that $1+a+b+3=4$, which implies that $a+b=0$. In addition to $2a-b=6$, this leads to $a=2, b=-2$.

Problem 4.8

(1) $\frac{3}{2}$. (2) $2(x-1)$. (3) $\frac{2}{(x-2)(x-4)}$. (4) $(x-2)$.

Problem 4.9

(1) $\frac{3x+1}{x(x+1)}$. (2) $\frac{7x+18}{(x+2)(x+3)}$.

(3) $\frac{5x-12}{(x-1)(x-2)}$. (4) $\frac{-x+27}{(x-2)(x+3)}$.

Problem 4.10

(1) $\frac{2x+7}{x+4}$. (2) $\frac{3}{(x+1)(2x-1)}$.

(3) $\frac{x-3}{2x+3}$. (4) $\frac{30}{(x-2)(x+3)}$.

Problem 4.11

(1) $\frac{2}{x+3}+\frac{3}{x-1}$. (2) $\frac{1}{x-2}+\frac{1}{x+1}$.

(3) $\frac{1}{x+3}+\frac{4}{x-2}$. (4) $\frac{2}{x-2}+\frac{5}{x-1}$.

Problem 4.12

(1) $\frac{1}{x+1}+\frac{2x+3}{x^2+3}$. (2) $\frac{2}{x+1}+\frac{2x+1}{x^2+x+4}$.

(3) $\frac{5}{x}-\frac{2}{x^2+3}$. (4) $\frac{5}{x-2}+\frac{-2x+1}{x^2+3x+1}$.

Problem 4.13

(1) $\frac{1}{(x-1)^2}$. (2) $\frac{3}{(x-1)}+\frac{8}{(x-1)^2}$.

(3) $\frac{4}{5(x-1)}-\frac{4}{5(x+4)}+\frac{1}{(x+4)^2}$. (4) $-\frac{1}{8(x+1)}+\frac{1}{8(x-3)}+\frac{1}{2(x-3)^2}$.

Problem 4.14

(1) $1+\frac{13}{2(x-1)}+\frac{15x+13}{2(x^2+1)}$. (2) $x+\frac{29}{5(x-2)}+\frac{15}{5(x+3)}$.

(3) $2x+\frac{4}{x+1}-\frac{9}{(x+1)^2}$. (4) $3+\frac{1}{5(x-2)}+\frac{-5x-64}{10(x^2+1)}$.

Problem 4.15

All figures are skipped.

(1) $f(x)=x^2-1$.

① The graph passes through the points $(1,0)$, $(-1,0)$, $(0,-1)$.

② Minimum point $(0,-1)$.

③ The slope of the tangent to the graph is $k=2x$.

(2) $f(x)=2x^2+5x+2$.

① The graph passes through the points $(-2,0)$, $\left(-\frac{1}{2},0\right)$, $(0,2)$.

② Minimum point $\left(-\frac{5}{4}, -\frac{9}{4}\right)$.

③ The slope of the tangent to the graph is $k=4x+5$.

(3) $f(x)=3x^3+11x^2+12x+4$.

① The graph passes through the points $(0,4)$, $\left(-\frac{3}{2},0\right)$, $(-1, 0)$, $(-2,0)$.

② Stationary points $(-0.82,-4.10)$ Minimum, $(-1.62,0.67)$ Maximum.

③ The slope of the tangent to the graph is $k=9x^2+22x+12$.

(4) $f(x)=2x^3-9x^2+10x-3$.

① The graph passes through the points $(0,-3)$, $\left(\frac{1}{2},0\right)$, $(1,0)$, $(2,0)$.

② Stationary points $(2.26,-3.28)$ Minimum, $(0.74, 0.28)$ Maximum.

③ The slope of the tangent to the graph is $k=6x^2-18x+10$.

Problem 4.16

(1) $A=2, B=-3, C=-11, D=6$. (2) $a=-14$.

(3) $\frac{2}{5(2x-1)}-\frac{1}{5(x+2)}$. (4) $(-\infty,-2)\cup(-1,3)$.

Problem 4.17

(1) $\frac{2}{x-1}-\frac{2}{2x+1}$. (2) $(-2,0),(0,-4)$.

Problem 4.18

(1) $b=3$, $c=5$. (2) $(x+1)(x+2)(2x-3)$.

(3) The figure is skipped.

Problem 4.19

(1) $\frac{x+1}{x-1}$. (2) $\frac{x+1}{x-1}$.

(3) Two graphs are same and symmetric about $y=x$.

(4) The figure is skipped.

Chapter 5 Trigonometry

The word "trigonometry" comes from the Greek words "trigon" (triangle) and "metron" (measurement), its initial meaning being "measuring the sides of a triangle", since trigonometry has been created to solve geometric problems in a triangle.

5.1 Angles and angle measurement

The concept of an angle is central to trigonometry. An angle as in Figure 5.1 has a **vertex** (or **origin**), O, and two **sides**, OA and OB, which extend indefinitely.

Angle notation

The angle can be denoted using the symbol $\angle$ followed by three capital letters describing the vertex and the sides, with the letter denoting the vertex in the middle. A notation for the angle pictured in Figure 5.1 would then be $\angle AOB$. Conversely, when there is no danger of confusion, the angle can be denoted just using the capital letter for its vertex. For instance, $\angle AOB$ could also be referred to as angle O.

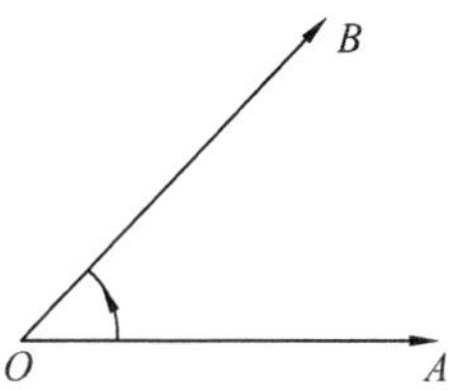

Figure 5.1 An angle $\angle AOB$ (or an angle O).

Angle measurement

To each angle, one may associate a number called the **measure** of the angle. In geometry, the measure of an angle is always between 0° and 360° and it is taken to be the smallest amount of rotation from the direction of one side of the angle to the direction of the other side, either clockwise or anticlockwise.

In trigonometry, although the measure of the angle is also the amount of rotation from the direction of one side of the angle to the other, the measure can be positive or negative as well and of any magnitude, and the angles are always measured anticlockwise. The measure of an angle can be larger than 360° when more than one rotation is involved, as seen in Figure 5.2 and can be negative is the measurement is done clockwise.

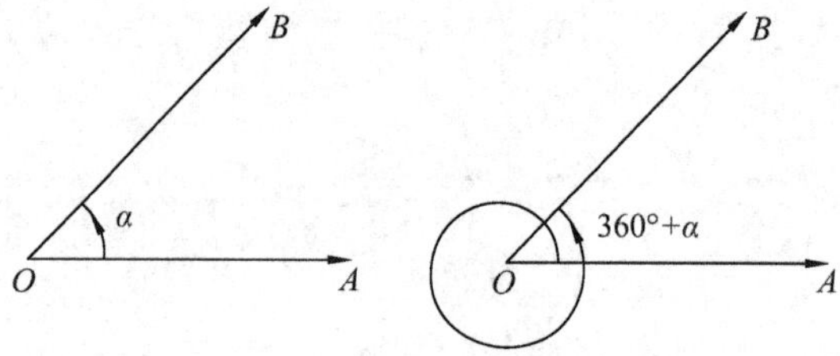

Figure 5.2 An angle whose measure exceeds 360°.

Measurement units

Angles are commonly measured using either the **sexagesimal system**, or the **radian** (circular) **measure.**

In the sexagesimal system, a right angle is divided into 90 equal parts, called **degrees.** A degree is denoted by °. For instance 45° means 45 degrees (half of a right angle). A degree is further divided into 60 parts, called **minutes.** A minute is denoted by ′ and is further divided into 60 **seconds** (″).

■ **Example 5.1** $45°20'27''$ means 45 degrees, 20 minutes and 35 seconds.

■

In the radian measure, as shown in Figure 5.3, the angle subtended at the center by an arc of circle whose length is equal to the radius of the circle is called a **radian** and is denoted by 1^cor 1 rad (although the ^{c} or rad are commonly omitted in writing).

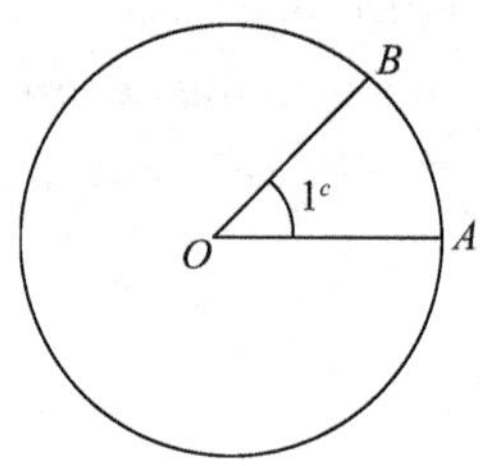

Figure 5.3 An angle whose measure is 1^c.

Since the length of the circumference is 2π times the radius, and angle of 180° subtends an arc of circle whose length is π times the radius. That is, its radian measure is π^c, or π rad or simply π. One then has the following conversion rule

$$\pi \text{ rad}=180°;$$

$$1°=\frac{\pi}{180};$$

$$1 \text{ rad}=\frac{180}{\pi},$$

which leads to the following conversion table for commonly used angles

0°	30°	45°	60°	90°	180°	360°
0	$\frac{\pi}{6}$	$\frac{\pi}{4}$	$\frac{\pi}{3}$	$\frac{\pi}{2}$	π	2π

5.2 Trigonometric functions

Definition 5.2.1—Trigonometric functions. In mathematics, the **trigonometric functions**, also called **circular functions** or **angle functions**, are functions of an angle. The most common trigonometric functions are the **sine** function (shortened as sin), the **cosine** function (shortened as cos) and the **tangent** function (shortened as tan).

5.2.1 Right triangle definitions

For acute angles, trigonometric functions are commonly defined as ratios of sides in a right triangle that contains the respective angle. That is, to define the trigonometric functions of an acute angle A, start with a right triangle that contains the angle A. The three sides of the triangle, indicated as such in Figure 5.4, are named as follows:

- the **hypotenuse** is the side opposite the right angle (the longest of all three sides);
- the **opposite** is the side opposite the angle we compute the trigonometric function for;
- the **adjacent** is the remaining side, which is near the angle but it is not the hypotenuse.

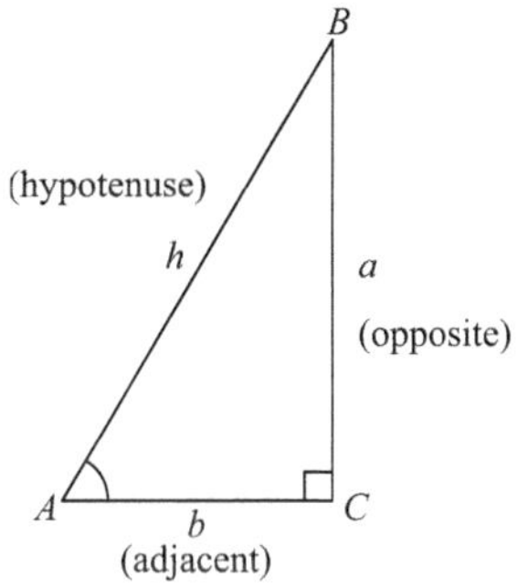

Figure 5.4 Right triangle *ABC*.

The trigonometric functions are then defined as indicated below.

$$\sin A=\frac{a}{h}=\frac{\text{opposite}}{\text{hypotenuse}};$$

$$\cos A=\frac{b}{h}=\frac{\text{adjacent}}{\text{hypotenuse}};$$

$$\tan A=\frac{a}{b}=\frac{\text{opposite}}{\text{adjacent}};$$

$$\cot A=\frac{b}{a}=\frac{\text{adjacent}}{\text{opposite}}=\frac{1}{\tan A};$$

$$\sec A=\frac{h}{b}=\frac{\text{hypotenuse}}{\text{adjacent}}=\frac{1}{\cos A};$$

$$\csc A = \frac{h}{a} = \frac{\text{hypotenuse}}{\text{opposite}} = \frac{1}{\sin A}.$$

The last three functions, the **cotangent** function (shortened as cot), the **secant** function (shortened as sec) and the **cosecant** function (shortened as csc) are the reciprocal of the more common tangent, cosine and sine functions, respectively.

Ⓡ One can also use the following naming convention, valid for all triangles, not necessarily right-angled: the opposite of an angle denoted with a capital letter is a side whose length is denoted with the corresponding small letter. Using this convention, $|AB| = c$, $|AC| = b$, $|BC| = a$.

■ **Example 5.2** Let ABC be a right triangle with its right angle in A. Find c knowing that $b=4$ and $\tan B=\frac{4}{5}$. For the angle B, c is the adjacent and b is the opposite. Then

$$\tan B = \frac{b}{c} \Rightarrow \frac{4}{5} = \frac{4}{c} \Rightarrow c = 5.$$

■

■ **Example 5.3** Let ABC be a right triangle with its right angle in A. Find c knowing that $h=10$ and $\sin C=\frac{1}{5}$.

For the angle C, c is the opposite. It follows that

$$\sin C = \frac{c}{h} \Rightarrow \frac{1}{5} = \frac{c}{10} \Rightarrow c = 2.$$

■

5.2.2 Unit circle definitions

The six trigonometric functions can also be defined using the unit circle, the circle of radius one centered at the origin. While the right-angled triangle definitions can only be given for angles between 0 and $\frac{\pi}{2}$ (see Figure 5.5), the definitions using the unit circle do not need to obey that restriction.

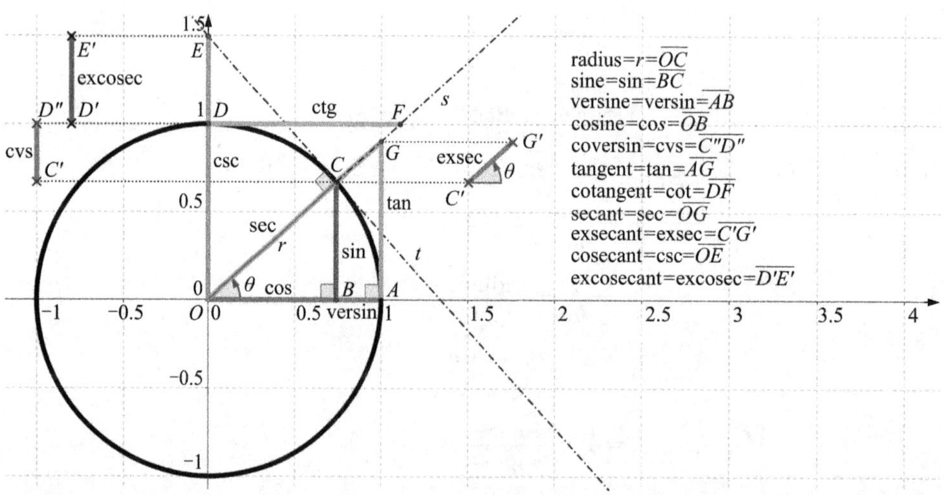

Figure 5.5 All trigonometric functions of the angle θ can be defined geometrically using the unit circle.

Let us draw a line through the center O of the circle, making an angle of θ with the positive half of the x-axis. Note that the measurement of the angle is done **anticlockwise**, which, again, is the right way in trigonometry. The line intersects the unit circle at a point C whose x-and y-coordinates are defined to be $\cos\theta$ and $\sin\theta$, respectively.

The periodicity of the sine and cosine functions

For angles greater than 2π, one simply continues to rotate around the circle, which leads to sin and cos being periodic functions with fundamental period 2π, that is

$\sin\theta = \sin(\theta + 2k\pi)$;

$\cos\theta = \cos(\theta + 2k\pi)$ for any angle θ and any integer k.

For angles less than 0, one measures the angle clockwise instead of anticlockwise.

■ **Example 5.4** If A, B are such that $A - B = 3\pi$, then prove that $\sin 2A = \sin 2B$.

Since $A = B + 3\pi$ and sin is a periodic function of period 2π, it follows that

$\sin 2A = \sin 2(B + 3\pi) = \sin(2B + 6\pi) = \sin(2B + 3\cdot 2\pi) = \sin 2B$.

■

Maximal and minimal values of the sine and cosine functions

From the unit circle definition, it follows that the values of the sine and cosine functions are at most 1 and at least -1, as they are the y-and x-coordinates, respectively, of a point moving on a circle of radius 1. That is,

$-1 \leqslant \sin\theta \leqslant 1$;

$-1 \leqslant \cos\theta \leqslant 1$ for any angle θ.

■ **Example 5.5** Is there an angle θ such that $\sin\theta = 2$? What about $\cos\theta = -3$?

The value 2 is too large (larger than 1). The value -3 is too small (smaller than -1). There are no such angles.

■

Values of the sine and cosine functions for common angles (Ⅰ)

From the unit circle definition, it follows that

$$\sin k\pi = 0 \text{ for any integer } k,$$

$$\cos\frac{(2k+1)\pi}{2} = 0 \text{ for any integer } k.$$

That is, the sine of any integer multiple of π and the cosine of any odd multiple of $\frac{\pi}{2}$ are 0.

■ **Example 5.6**

$$\sin\pi = \sin 2\pi = \cos\frac{3\pi}{2} = \cos\frac{7\pi}{2} = 0.$$

■

The sign of the sine and cosine functions

Analyzing the signs of x-and y- coordinates in each quadrant together with the unit circle, one obtains that

$\sin\theta \geqslant 0$, $\theta \in [0, \pi]$;

$\sin\theta \leqslant 0$, $\theta \in [\pi, 2\pi]$;

$\cos\theta \geqslant 0$, $\theta \in \left[0, \frac{\pi}{2}\right] \cup \left[\frac{3\pi}{2}, 2\pi\right]$;

$\cos\theta \leqslant 0$, $\theta \in \left[\frac{\pi}{2}, \frac{3\pi}{2}\right]$.

® Particularly, if A is an angle in a triangle ABC, $\sin A$ is positive regardless of A being an acute, right or obtuse angle. However, $\cos A$ is positive if A is an acute angle, zero if A is a right angle, and negative if A is an obtuse angle.

■ **Example 5.7** Prove that $\sin 183° < 0$ and $\cos 183° < 0$.

Let us draw a line through the center O of the circle, making an angle of 183° with the positive half of the x-axis. The line intersects the unit circle at a point C whose x-and y-coordinates are defined to be $\cos 183°$ and $\sin 183°$, respectively. Since C is in the third quadrant, in which both x and y are negative, $\sin 183° < 0$ and $\cos 183° < 0$.

■

■ **Example 5.8** In a triangle ABC, the angles A, B are such that $\cos A = -\frac{1}{2}$, $\cos B = \frac{\sqrt{3}}{2}$. Which one is larger?

Since $\cos A < 0$ and $\cos B > 0$, A is obtuse and B is acute. The angle A is larger.

■

Values of the sine and cosine functions for common angles (II)

It is easy to remember the values of sin and cos for a few common angles, namely for 0°, 30°, 45°, 60°, 90°, with the help of the table below.

x	0°	30°	45°	60°	90°
$\sin x$	$\frac{\sqrt{0}}{2}$	$\frac{\sqrt{1}}{2}$	$\frac{\sqrt{2}}{2}$	$\frac{\sqrt{3}}{2}$	$\frac{\sqrt{4}}{2}$
$\cos x$	$\frac{\sqrt{4}}{2}$	$\frac{\sqrt{3}}{2}$	$\frac{\sqrt{2}}{2}$	$\frac{\sqrt{1}}{2}$	$\frac{\sqrt{0}}{2}$

The mnemonics is quite simple: on the sin line, put 0, 1, 2, 3, 4, and on the cos line, put the same numbers, but in decreasing order. Then take square roots and divide each square root by 2 afterwards. Put then the result in the simplest form.

■ **Example 5.9**

$$\sin 90° = \frac{\sqrt{4}}{2} = 1, \ \cos 90° = \frac{\sqrt{0}}{2} = 0.$$

■

■ **Example 5.10** Find $\sin\left(\frac{\pi}{3} + 100\pi\right)$ and $\cos\left(\frac{101\pi}{2}\right)$.

Since the fundamental period for the sine function is 2π, 100π means 50 periods. Since the addition or removal of an integer number of periods from the argument of a periodic function does not modify the value of that function, it follows that

$$\sin\left(\frac{\pi}{3}+100\pi\right)=\sin\left(\frac{\pi}{3}+50\cdot 2\pi\right)=\sin\left(\frac{\pi}{3}\right)=\frac{\sqrt{3}}{2}.$$

For the second value, one should see first how many periods does $\frac{101\pi}{2}$ contain. Since

$$\frac{101\pi}{2}=\frac{\pi}{2}+50\pi=\frac{\pi}{2}+25\cdot 2\pi,$$

it follows that

$$\cos\left(\frac{101\pi}{2}\right)=\cos\left(\frac{\pi}{2}+25\cdot 2\pi\right)=\cos\left(\frac{\pi}{2}\right)=0.$$

■

From the above table, one may also deduce the corresponding values for the tangent and cotangent function.

■ **Example 5.11**

$$\tan 45^\circ=\frac{\sin 45^\circ}{\cos 45^\circ}=\frac{\frac{\sqrt{2}}{2}}{\frac{\sqrt{2}}{2}}=1,\ \tan 60^\circ=\frac{\sin 60^\circ}{\cos 60^\circ}=\frac{\frac{\sqrt{3}}{2}}{\frac{1}{2}}=\frac{\sqrt{3}}{2}\cdot\frac{2}{1}=\sqrt{3}.$$

■

5.3 Equations and identities

The equation of the unit circle is

$$x^2+y^2=1.$$

Let us use this equation to establish several trigonometric identities of practical use.

5.3.1 Pythagorean trigonometric identities

From the figure above, OC is the radius of the circle and the hypotenuse of the triangle OBC. Denoting OC as r, OB as x and BC as y, we have the following:

$$r^2=x^2+y^2.$$

Dividing both sides by r^2, we get

$$\frac{r^2}{r^2}=\frac{x^2}{r^2}+\frac{y^2}{r^2},$$

that is $\left(\frac{x}{r}\right)^2+\left(\frac{y}{r}\right)^2=1$ which is the same as

$$\sin^2\theta+\cos^2\theta=1.$$

This equality is called **the fundamental Pythagorean trigonometric identity.**

Ⓡ Note that $\sin^2\theta$ and $\cos^2\theta$ are the usual shorthands for $(\sin\theta)^2$ and $(\cos\theta)^2$, respectively.

■ **Example 5.12** If $\theta\in(0,\frac{\pi}{2})$ and $\sin\theta=\frac{4}{5}$, then find $\cos\theta$.

Since $\sin^2\theta+\cos^2\theta=1$, it follows that $\frac{16}{25}+\cos^2\theta=1$, which implies that $\cos^2\theta=\frac{9}{25}$. The possible values of $\cos\theta$ are then $\frac{3}{5}$ and $-\frac{3}{5}$. Since $\theta\in\left(0, \frac{\pi}{2}\right)$, $\cos\theta$ is positive, and then $\cos\theta=\frac{3}{5}$.

■ **Example 5.13** Is there any angle θ such that $\sin\theta=0.8$ and $\cos\theta=0.5$?

Both $\cos\theta$ and $\sin\theta$ are given. We verify the validity of the fundamental Pythagorean trigonometric identity and obtain

$$\sin^2\theta+\cos^2\theta=0.64+0.25=0.89\neq 1.$$

The fundamental Pythagorean trigonometric identity is not verified and, consequently, there is no such angle.

■

Dividing the fundamental Pythagorean trigonometric identity by $\cos^2\theta$ and $\sin^2\theta$, respectively, we have

$$\sec^2\theta=1+\tan^2\theta;$$

$$\csc^2\theta=1+\cot^2\theta.$$

These identities are also called **Pythagorean trigonometric identities.**

Ⓡ The above identities can also be written as

$$1+\tan^2\theta=\frac{1}{\cos^2\theta};$$

$$1+\cot^2\theta=\frac{1}{\sin^2\theta}.$$

Notice the mnemonics: adding 1 to the square of the tangent or cotangent function leads to replacing the numerator of that square with 1.

Using the Pythagorean trigonometric identities together with sign considerations, it is possible to express any of the six basic trigonometric functions in terms of any other.

■ **Example 5.14** Find $\sin^2\theta$ knowing that $\cot\theta=2$.

Since $1+\cot^2\theta=\frac{1}{\sin^2\theta}$, it follows that

$$1+2^2=\frac{1}{\sin^2\theta}\Rightarrow\sin^2\theta=\frac{1}{5}.$$

■

5.3.2 Addition and subtraction formulae

Addition and subtraction formulae for the sine and cosine functions

The addition and subtraction formulae for sin and cos are given by

$$\sin(A+B)=\sin A\cos B+\cos A\sin B;$$

$$\cos(A+B)=\cos A\cos B-\sin A\sin B;$$

$$\sin(A-B)=\sin A\cos B-\cos A\sin B;$$

$$\cos(A-B)=\cos A\cos B+\sin A\sin B.$$

Ⓡ Notice the mnemonics: sin starts with the letter s (as it does sign), then it preserves the sign. As a result, $\sin(A+B)$ has a $+$ between the terms in the right-hand side, while $\sin(A-B)$ has a $-$ between them. Also, cos does not start with an s, then it does not preserve the sign. As a result, $\cos(A+B)$ has a $-$ between the terms in the right-hand side, not a $+$, while $\cos(A-B)$ has a $+$ between them, not a $-$. Since sin preserves the sign, it "mixes" the functions (both terms in the right-hand side of $\sin(A+B)$ and $\sin(A-B)$ mix together a sin and a cos).

Since cos "mixes" the sign, it preserves the functions (both terms in the right-hand side of $\cos(A+B)$ and $\cos(A-B)$ do not mix sin and cos, but are rather of type coscos and sinsin).

■ **Example 5.15** Evaluate $\sin\left(\frac{\pi}{6}+\frac{\pi}{4}\right)$.

Applying the addition formula for the sine function, we see that

$$\sin\left(\frac{\pi}{6}+\frac{\pi}{4}\right)=\sin\frac{\pi}{6}\cos\frac{\pi}{4}+\cos\frac{\pi}{6}\sin\frac{\pi}{4}=\frac{1}{2}\cdot\frac{\sqrt{2}}{2}+\frac{\sqrt{3}}{2}\cdot\frac{\sqrt{2}}{2}=\frac{\sqrt{6}+\sqrt{2}}{4}.$$

■

■ **Example 5.16** Evaluate

(1) $\sin 10^\circ\cos 20^\circ+\cos 10^\circ\sin 20^\circ$;

(2) $\cos 11^\circ\cos 49^\circ-\sin 11^\circ\sin 49^\circ$.

(1) This is the expansion of a sin (mixed functions)

$$\sin 10^\circ\cos 20^\circ+\cos 10^\circ\sin 20^\circ=\sin(10^\circ+20^\circ)=\sin(30^\circ)=\frac{1}{2}.$$

(2) This is the expansion of a cos (non-mixed functions)

$$\cos 11^\circ\cos 49^\circ-\sin 11^\circ\sin 49^\circ=\cos(11^\circ+49^\circ)=\cos(60^\circ)=\frac{1}{2}.$$

■

Exercise 5.1 Prove that

$\sin(A+B)\sin(A-B)=\sin^2 A-\sin^2 B$ for any angles A and B.

■

Proof Applying the addition and subtraction formulae for the sine function, we see that

$$\begin{aligned}\sin(A+B)\sin(A-B)&=(\sin A\cos B+\cos A\sin B)\cdot(\sin A\cos B-\cos A\sin B)\\&=(\sin A\cos B)^2-(\cos A\sin B)^2\\&=\sin^2 A\cos^2 B-\cos^2 A\sin^2 B.\end{aligned}$$

From the fundamental Pythagorean identity, $\cos^2 B=1-\sin^2 B$ and $\cos^2 A=1-\sin^2 A$. Substituting these equalities, one obtains

$$\begin{aligned}\sin(A+B)\sin(A-B)&=\sin^2 A(1-\sin^2 B)-(1-\sin^2 A)\sin^2 B\\&=\sin^2 A-\sin^2 A\sin^2 B-\sin^2 B+\sin^2 A\sin^2 B\\&=\sin^2 A-\sin^2 B.\end{aligned}$$

■

Exercise 5.2 Prove that

$$\sin A\sin(B-C)+\sin B\sin(C-A)+\sin C\sin(A-B)=0.$$

■

Proof Applying the subtraction formulae, one obtains

$$\sin(B-C)=\sin B\cos C-\cos B\sin C,$$
$$\sin(C-A)=\sin C\cos A-\cos C\sin A,$$
$$\sin(A-B)=\sin A\cos B-\cos A\sin B.$$

Multiplying by $\sin A$, $\sin B$, $\sin C$, respectively, we obtain

$$\sin A\sin(B-C)=\sin A\sin B\cos C-\sin A\cos B\sin C,$$
$$\sin B\sin(C-A)=\cos A\sin B\sin C-\sin A\sin B\cos C,$$
$$\sin C\sin(A-B)=\sin A\cos B\sin C-\cos A\sin B\sin C,$$

which imply that the required identity holds.

■

Addition and subtraction formulae for the tangent function

The addition and subtraction formulae for tan are given by

$$\tan(A+B)=\frac{\tan A+\tan B}{1-\tan A\tan B};$$

$$\tan(A-B)=\frac{\tan A-\tan B}{1+\tan A\tan B}.$$

Ⓡ Notice the mnemonics: the numerator preserves the original sign, while the denominator contains the opposite sign.

■ **Example 5.17** Evaluate $\tan 105°$.

$$\tan 105°=\tan(60°+45°)$$
$$=\frac{\tan 60°+\tan 45°}{1-\tan 60°\tan 45°} \quad \text{(addition formula)}$$
$$=\frac{\sqrt{3}+1}{1-\sqrt{3}}.$$

■

■ **Example 5.18** Find $\tan(A+B)$, knowing that $\cot A=\frac{1}{2}$ and $\cot B=\frac{1}{3}$.

Since $\cot A=\frac{1}{2}$ and $\cot B=\frac{1}{3}$, it follows that $\tan A=\frac{1}{\cot A}=\frac{1}{\frac{1}{2}}=2$, $\tan B=\frac{1}{\cot B}=\frac{1}{\frac{1}{3}}=3$. Then

$$\tan(A+B)=\frac{\tan A+\tan B}{1-\tan A\tan B}=\frac{2+3}{1-2\cdot 3}=-1.$$

■

■ **Example 5.19** If $\tan\theta=\frac{1}{2}$, then find $\tan\left(\theta+\frac{\pi}{4}\right)$.

Using the addition formula for the tangent function, one obtains

$$\tan\left(\theta+\frac{\pi}{4}\right)=\frac{\tan\theta+\tan\frac{\pi}{4}}{1-\tan\theta\tan\frac{\pi}{4}}.$$

Since $\tan\frac{\pi}{4}=1$ and $\tan\theta=\frac{1}{2}$, one has

$$\tan\left(\theta+\frac{\pi}{4}\right)=\frac{\frac{1}{2}+1}{1-\frac{1}{2}}=\frac{\frac{3}{2}}{\frac{1}{2}}=3.$$

■

Parity of the sine, cosine and tangent function

Substituting $A=0$ in the formula for $\sin(A-B)$, we obtain

$$\sin(-B)=-\sin B,$$

consistent with the mnemonics that sin preserves the sign (i.e., it does not make the − disappear). That is, **the sine function is an odd function.**

Substituting $A=0$ in the formula for $\cos(A-B)$, we obtain

$$\cos(-B)=\cos B,$$

consistent with the mnemonics that cos does not preserve the sign (i.e., it makes the − disappear). That is, **the cosine function is an even function.**

Finally, substituting $A=0$ in the formula for $\tan(A-B)$, we obtain

$$\tan(-B)=-\tan B.$$

That is, **the tangent function is an odd function.**

■ **Example 5.20** Find $\sin(-30°)$, $\cos(-45°)$, $\tan(-60°)$.

Using the parity (or imparity) formulae, one sees that

$$\sin(-30°)=-\sin(30°)=-\frac{1}{2};$$

$$\cos(-45°)=\cos(45°)=\frac{\sqrt{2}}{2};$$

$$\tan(-60°)=-\tan(60°)=-\sqrt{3}.$$

■

5.3.3 Functions of the complementary angle

Definition 5.3.1—Complementary angle. The complement of an angle B is the angle $\frac{\pi}{2}-B$ (the angle to which B is added to obtain $\frac{\pi}{2}$). For instance, in a right triangle, the acute angles are complementary to each other.

Substituting $A=\frac{\pi}{2}$ in the formulae for $\sin(A-B)$, $\cos(A-B)$, $\tan(A-B)$, we have

$$\sin\left(\frac{\pi}{2}-B\right)=\cos B;$$

$$\cos\left(\frac{\pi}{2}-B\right)=\sin B;$$

$$\tan\left(\frac{\pi}{2}-B\right)=\cot B;$$

Ⓡ Notice the mnemonics: pairing sin with cos and tan with cot, the value of a function of the complementary angle is equal to the value of the pairing function for the initial angle.

■ **Example 5.21** Prove that $\sin 53°=\cos 37°$ and $\tan 1°=\cot 89°$.

The angles of 53° and 37° are complementary, and so are also those of 1° and 89°. Consequently,

$$\sin 53°=\sin(90°-37°)=\cos 37°;$$
$$\tan 1°=\tan(90°-89°)=\cot 89°.$$

■

Exercise 5.3 Find the value of

$$P=\tan 1°\ \tan 2°\ \tan 3°\ \cdots\tan 89°.$$

■

Solution The idea is to pair the tangent of an angle with the tangent of their complement. One sees that

$$\begin{aligned}P^2=P\times P&=\tan 1°\ \tan 2°\ \tan 3°\cdots\tan 89°\times\tan 1°\ \tan 2°\ \tan 3°\cdots\tan 89°\\&=\tan 1°\ \tan 2°\ \tan 3°\cdots\tan 89°\times\tan 89°\ \tan 88°\ \tan 87°\cdots\tan 1°\\&=(\tan 1°\ \tan 89°)\times(\tan 2°\ \tan 88°)\times(\tan 3°\ \tan 87°)\times\cdots\times(\tan 89°\ \tan 1°).\end{aligned}$$

However, the product of tangents in each pair is 1, since

$$\tan(k°)\ \tan(90°-k°)=\tan(k°)\ \cot(k°)=\tan k°\frac{1}{\tan(k°)}=1.$$

Consequently, $P^2=1$, and then $P=1$, since all factors of P are positive.

■

5.3.4 Functions of the supplementary angle

Definition 5.3.2—Supplementary angle. The supplement of an angle B is the angle $\pi-B$(the angle to which B is added to obtain π).

Substituting $A=\pi$ in the formulae for $\sin(A-B)$, $\cos(A-B)$, $\tan(A-B)$, we have

$$\sin(\pi-B)=\sin B;$$
$$\cos(\pi-B)=-\cos B;$$
$$\tan(\pi-B)=-\tan B.$$

Ⓡ The mnemonics is actually clearer if we substitute $A=\pi$ in the formulae for $\sin(A+B)$, $\cos(A+B)$, $\tan(A+B)$. We have

$$\sin(\pi+B)=-\sin B;$$
$$\cos(\pi+B)=-\cos B;$$
$$\tan(\pi+B)=\tan B.$$

That is, the removal of π from the argument of a sin or a cos puts a $-$ in front of the value, while the removal of π from the argument of a tan has no effect. Then these formulae can be thought as "removal of π" formulae. It follows that

$$\sin(\pi-B)=-\sin(-B)=\sin B;$$
$$\cos(\pi-B)=-\cos(-B)=-\cos B;$$
$$\tan(\pi-B)=\tan(-B)=-\tan B,$$

since the sine and tangent functions are odd, while the cosine function is even.

Ⓡ As seen from

$$\tan(\pi+B)=\tan B,$$

the fundamental period of the tangent function is π, not 2π.

■ **Example 5.22** Find

(1) $\sin(135°)$;

(2) $\cos(210°)$;

(3) $\tan(120°)$.

All angles are larger 90°. We see that

(1) $\sin(135°)=\sin(180°-45°)=-\sin(-45°)=\sin(45°)=\frac{\sqrt{2}}{2}$;

(2) $\cos(210°)=\cos(180°+30°)=-\cos(30°)=\frac{\sqrt{3}}{2}$;

(3) $\tan(120°)=\tan(180°-60°)=\tan(-60°)=-\tan(60°)=-\sqrt{3}$.

■

■ **Example 5.23** If A, B are angles such that $A-B=\pi$, prove that $\cos A\cos B\leqslant 0$.

Since $A=B+\pi$, it follows that

$$\begin{aligned}\cos A\cos B&=\cos(B+\pi)\cos B\\&=-\cos B\cdot\cos B\\&=-\cos^2 B\leqslant 0.\end{aligned}$$

■

Exercise 5.4 Prove that

$$\sin(\pi+\theta)+\sin(\pi-\theta)=0 \text{ for any angle } \theta.$$

■

Proof By "removing" π, one has

$$\sin(\pi+\theta)=-\sin\theta,$$
$$\sin(\pi-\theta)=-\sin(-\theta)=\sin\theta,$$

and then

$$\sin(\pi+\theta)+\sin(\pi-\theta)=-\sin\theta+\sin\theta=0.$$

■

5.3.5 Double angle formulae

Substituting $A=B$ in addition formulae, we have

$$\sin 2A=2\sin A\cos A;$$
$$\cos 2A=\cos^2 A-\sin^2 A$$

$$=2\cos^2 A-1$$
$$=1-2\sin^2 A;$$
$$\tan 2A=\frac{2\tan A}{1-\tan^2 A}.$$

These formulae can be used to express trigonometric functions of the double angle $2A$ when trigonometric functions of the single angle A are known.

■ **Example 5.24** Find $\tan 2\theta$, knowing that $\cot \theta=3$.

We find $\tan \theta$ first. Since $\cot \theta=3$, $\tan \theta=\frac{1}{\cot \theta}=\frac{1}{3}$. Using the double angle formula for the tangent function, we then obtain

$$\tan 2\theta=\frac{2\tan \theta}{1-\tan^2 \theta}=\frac{2\cdot\frac{1}{3}}{1-\left(\frac{1}{3}\right)^2}=\frac{3}{4}.$$

■

■ **Example 5.25** Prove that $(\sin \theta+\cos \theta)^2=1+\sin 2\theta$.

$(\sin \theta+\cos \theta)^2=\sin^2 \theta+2\sin \theta \cos \theta+\cos^2 \theta$

$=1+2\sin \theta \cos \theta$ (fundamental Pythagorean identity)

$=1+\sin 2\theta$ (double angle formula)

■

■ **Example 5.26** Prove that $\cos^4 \theta-\sin^4 \theta=\cos 2\theta$.

$\cos^4 \theta-\sin^4 \theta=(\cos^2 \theta)^2-(\sin^2 \theta)^2$

$=(\cos^2 \theta-\sin^2 \theta)(\cos^2 \theta+\sin^2 \theta)$ (difference of squares formula)

$=\cos^2 \theta-\sin^2 \theta$ (fundamental Pythagorean identity)

$=\cos 2\theta$ (double angle formula)

■

Rearranging the double angle formula for the cosine function, we obtain the following formulae, which are sometimes useful to remove squares from trigonometric expressions

$$\cos^2 A=\frac{1+\cos 2A}{2};$$
$$2\sin^2 A=\frac{1-\cos 2A}{2}.$$

■ **Example 5.27** Prove that $1+\sin \theta=2\cos^2\left(\frac{\pi}{4}-\frac{\theta}{2}\right)$.

$2\cos^2\left(\frac{\pi}{4}-\frac{\theta}{2}\right)=2\,\frac{1+\cos 2\left(\frac{\pi}{4}-\frac{\theta}{2}\right)}{2}$ (double angle formula)

$=1+\cos\left(\frac{\pi}{2}-\theta\right)$

$=1+\sin \theta.$ (complementary angle formula)

■

Exercise 5.5 Prove that $\cos^4\frac{\pi}{12}+\cos^4\frac{7\pi}{12}=\frac{7}{8}$.

■

Proof

$$\cos^4\frac{\pi}{12}+\cos^4\frac{7\pi}{12}=\left(\cos^2\frac{\pi}{12}\right)^2+\left(\cos^2\frac{7\pi}{12}\right)^2$$

$$=\left(\frac{1+\cos 2\cdot\frac{\pi}{12}}{2}\right)^2+\left(\frac{1+\cos 2\cdot\frac{7\pi}{12}}{2}\right)^2 \qquad \text{(double angle formula)}$$

$$=\left(\frac{1+\cos\frac{\pi}{6}}{2}\right)^2+\left(\frac{1+\cos\frac{7\pi}{6}}{2}\right)^2.$$

Since $\cos\frac{7\pi}{6}=\cos\left(\pi+\frac{\pi}{6}\right)=-\cos\frac{\pi}{6}=-\frac{\sqrt{3}}{2}$, we see that

$$\cos^4\frac{\pi}{12}+\cos^4\frac{7\pi}{12}=\left(\frac{1+\frac{\sqrt{3}}{2}}{2}\right)^2+\left(\frac{1-\frac{\sqrt{3}}{2}}{2}\right)^2$$

$$=\frac{1+2\frac{\sqrt{3}}{2}+\left(\frac{\sqrt{3}}{2}\right)^2}{4}+\frac{1-2\frac{\sqrt{3}}{2}+\left(\frac{\sqrt{3}}{2}\right)^2}{4}$$

$$=\frac{2+2\cdot\frac{3}{4}}{4}$$

$$=\frac{7}{8}.$$

■

5.3.6 Half angle formulae

For solving trigonometric equations, simplifying expressions and for certain other more advances considerations, it can be useful to express trigonometric functions in terms of the tangent of the half-angle, as seen below

$$\tan A=\frac{2\tan\frac{A}{2}}{1-\tan^2\frac{A}{2}};$$

$$\sin A=\frac{2\tan\frac{A}{2}}{1+\tan^2\frac{A}{2}};$$

$$\cos A=\frac{1-\tan^2\frac{A}{2}}{1+\tan^2\frac{A}{2}}.$$

■ **Example 5.28**　Prove that $\frac{\sin 2\theta}{\cos\theta+\cos^2\theta}=2\tan\frac{\theta}{2}$.

$$\frac{\sin 2\theta}{\cos\theta+\cos^2\theta}=\frac{2\sin\theta\cos\theta}{\cos\theta(1+\cos\theta)} \qquad \text{(double angle formula)}$$

$$=\frac{2\sin\theta}{1+\cos\theta}$$

$$=\frac{2\,\dfrac{2\tan\dfrac{\theta}{2}}{1+\tan^2\dfrac{\theta}{2}}}{1+\dfrac{1-\tan^2\dfrac{\theta}{2}}{1+\tan^2\dfrac{\theta}{2}}} \qquad \text{(half angle formula)}$$

$$=\frac{\dfrac{4\tan\dfrac{\theta}{2}}{1+\tan^2\dfrac{\theta}{2}}}{\dfrac{2}{1+\tan^2\dfrac{\theta}{2}}}$$

$$=2\tan\frac{\theta}{2}.$$

■

Triple-angle formulae

Substituting $B=2A$ in the addition formulae, one obtains the following triple-angle formulae:

$$\sin 3A=3\sin A-4\sin^3 A;$$

$$\cos 3A=4\cos^3 A-3\cos A;$$

$$\tan 3A=\frac{3\tan A-\tan^3 A}{1-3\tan^2 A}.$$

Exercise 5.6 Prove that

$$\cos^3 x\sin 3x+\sin^3 x\cos 3x=\frac{3}{4}\sin 4x.$$

■

Proof

$$\cos^3 x\sin 3x+\sin^3 x\cos 3x=\cos^3 x(3\sin x-4\sin^3 x)+\sin^3 x(4\cos^3 x-3\cos x) \qquad \text{(triple angle formulae)}$$

$$=3\cos^3 x\sin x-3\sin^3 x\cos x$$

$$=3\cos x\sin x(\cos^2 x-\sin^2 x)$$

$$=\frac{3}{2}\sin 2x\cos 2x \qquad \text{(double angle formulae)}$$

$$=\frac{3}{4}\sin 4x. \qquad \text{(double angle formula)}$$

■

5.4 The cosine and sine laws

5.4.1 Law of cosines

In trigonometry, the law of cosines relates the lengths of the sides of a triangle to the cosine of one of its angles. Using the notations of Figure 5.6 (the angles α, β, and γ are respectively opposite the sides a, b and c), the law of cosines states that

$$\cos\alpha=\frac{b^2+c^2-a^2}{2bc} \text{ or } a^2=b^2+c^2-2bc\cos\alpha;$$

$$\cos\beta=\frac{a^2+c^2-b^2}{2ac} \text{ or } b^2=a^2+c^2-2ac\cos\beta;$$

$$\cos\gamma=\frac{a^2+b^2-c^2}{2ab} \text{ or } c^2=a^2+b^2-2ab\cos\gamma.$$

Ⓡ Notice the mnemonics: the equality

$$a^2=b^2+c^2-2bc\cos\alpha$$

(for instance) is similar, for the most part, to the Pythagorean theorem, written for the side of length a as the hypotenuse, except for the double product of the other sides multiplied by the cosine of the angle opposite the side of length a, which is subtracted from the right-hand side.

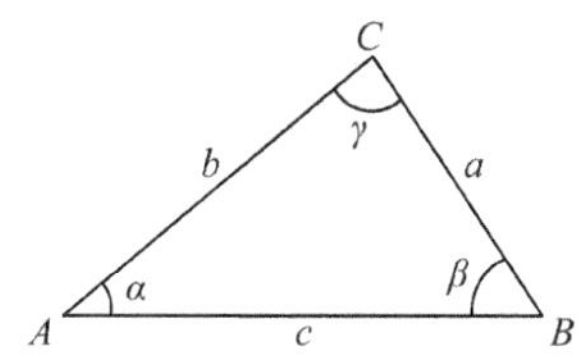

Figure 5.6 The triangle *ABC*. The angles α, β and γ are respectively opposite the sides a, b and c.

Proofs

Using the distance formula

Consider a triangle with sides of length a, b, c, where θ is the measure of the angle opposite the side of length c. This triangle can be placed on the Cartesian coordinate system by plotting the points, as shown in Figure 5.7, $A(b\cos\theta, b\sin\theta)$, $B(a,0)$ and $C(0,0)$. By the distance formula, we have

$$c=\sqrt{(a-b\cos\theta)^2+(0-b\sin\theta)^2},$$

which implies that

$$c^2=(a-b\cos\theta)^2+(0-b\sin\theta)^2=a^2+b^2-2ab\cos\theta.$$

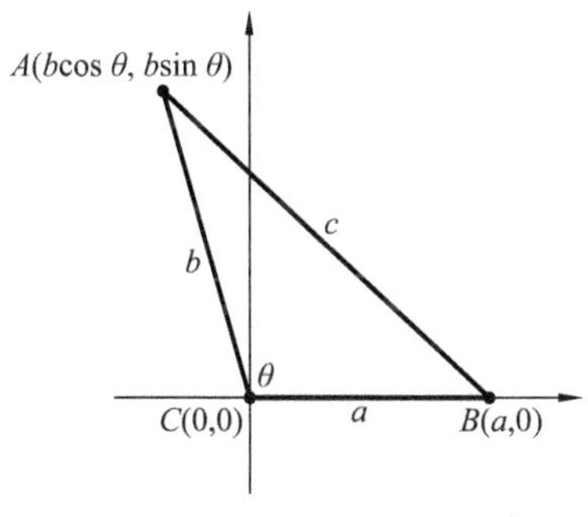

Figure 5.7 The triangle *ABC*. The angles *A*, *B* and *C* are respectively to opposite the sides a, b and c.

Ⓡ An advantage of this proof is that it does not require the consideration of different cases for when the triangle is acute or right or obtuse.

Using trigonometry

As shown in Figure 5.8, drop the perpendicular onto the side c to get

$$c=a\cos\beta+b\cos\alpha.$$

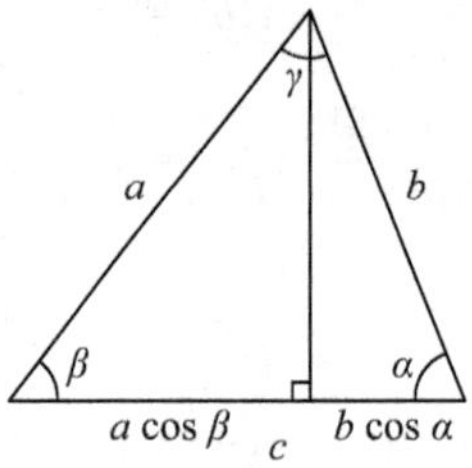

Figure 5.8 An acute triangle with perpendicular.

Multiply through by c to get

$$c^2=ac\cos\beta+bc\cos\alpha.$$

Similarly, by considering the other perpendiculars we obtain

$$a^2=ac\cos\beta+ab\cos\gamma$$

and

$$b^2=bc\cos\alpha+ab\cos\gamma.$$

Finally, we get

$$c^2=a^2+b^2-2ab\cos\gamma.$$

Ⓡ This proof uses trigonometry in that it treats the cosines of the various angles as quantities in their own right. It uses the fact that the cosine of an angle expresses the relation between the two sides enclosing that angle in **any** right triangle.

Using the Pythagorean theorem

As shown in Figure 5.9, which is the case of C being an obtuse angle, we apply the Pythagorean theorem to each of the two right triangles AHB and CHB. Using d to denote the line segment CH and h for the height BH, triangle AHB gives us

$$c^2=(b+d)^2+h^2,$$

and triangle CHB gives

$$d^2+h^2=a^2.$$

Also, $d=a\cos(\pi-\angle ACB)=a\cos(\pi-\gamma)=-a\cos\gamma$. Hence, we obtain

$$c^2=a^2+b^2-2ab\cos\gamma.$$

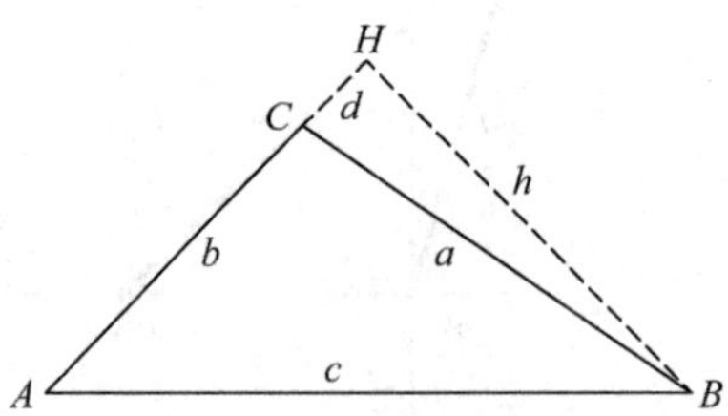

Figure 5.9 Obtuse triangle *ABC* with heigh *BH*.

As shown in Figure 5.10, which is the case of C being an acute angle, by using the right triangle on the left hand side, we obtain

$$\begin{aligned} c^2 &= (b - a\cos\gamma)^2 + (a\sin\gamma)^2 \\ &= b^2 - 2ab\cos\gamma + a^2\cos^2\gamma + a^2\sin^2\gamma \\ &= b^2 + a^2 - 2ab\cos\gamma. \end{aligned}$$

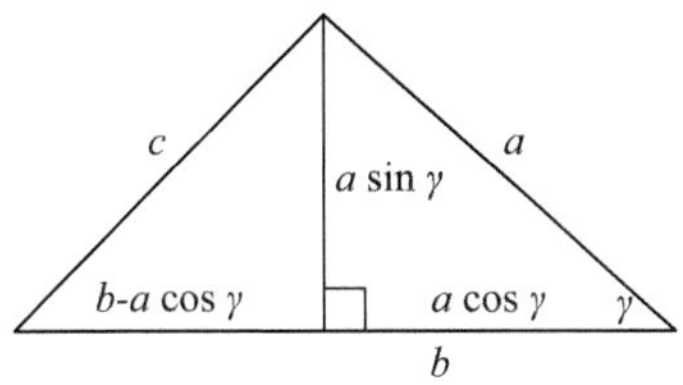

Figure 5.10 An acute triangle.

■ **Example 5.29** Let ABC be a triangle with $a=30$, $b=24$, $c=15$. Prove that one of its angles is obtuse. The largest angle is expected to be opposite the largest side. We then prove that the angle A is obtuse.

$$\cos A = \frac{b^2 + c^2 - a^2}{2bc} = \frac{24^2 + 15^2 - 30^2}{720} = -\frac{99}{720} < 0.$$

Since $\cos A < 0$, it follows that A is obtuse.

■

■ **Example 5.30** Using the data in Figure 5.11, find the length of side x by means of the law of cosines formula.

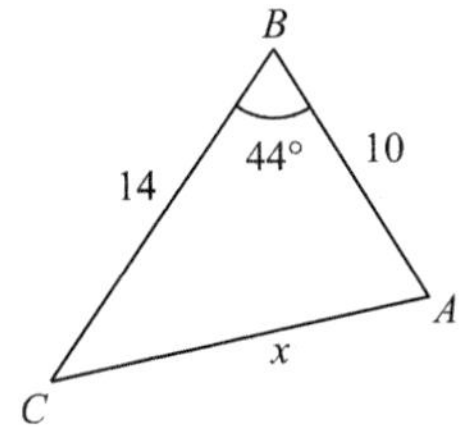

Figure 5.11 A triangle *ABC*.

$b^2 = a^2 + c^2 - 2ac\cos 44°$

$x^2 = 14^2 + 10^2 - 2 \cdot 14 \cdot 10 \cdot \cos 44°$

$x^2 = 296 - 280\cos 44°$

$x^2 \approx 94.585$

$x \approx 9.725\,475.$

■

■ **Example 5.31** Using the data in Figure 5.12, find the measure of $\angle A$ by means of the law of cosines formula.

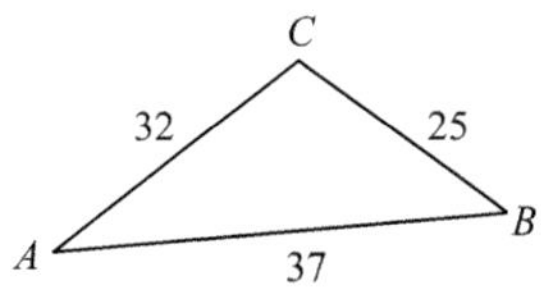

Figure 5.12 A triangle *ABC*.

$$a^2=b^2+c^2-2bc\cos A$$
$$25^2=32^2+37^2-2\cdot 32\cdot 37\cdot \cos A$$
$$625=2\,393-2\,368\cos A$$
$$A\approx\cos^{-1}(0.746\,6)$$
$$A\approx 41.70^\circ.$$

■

Exercise 5.7 Prove that, in any triangle ABC, $b\cos C+c\cos B=a$. ■

Proof Since the identity to be proven contains two cosine functions and two side lengths, we replace the cosine functions using the law of cosines, so that the result will contains only side lengths. We obtain

$$\begin{aligned} b\cos C+c\cos B &= b\frac{a^2+b^2-c^2}{2ab}+c\frac{a^2+c^2-b^2}{2ac} \\ &= \frac{a^2+b^2-c^2}{2a}+\frac{a^2+c^2-b^2}{2a} \\ &= \frac{2a^2}{2a} \\ &= a. \end{aligned}$$

■

5.4.2 Law of sines

In trigonometry, the law of sines is a formula relating the lengths of the sides of a triangle to the sines of its angles. According to this formula,

$$\frac{a}{\sin A}=\frac{b}{\sin B}=\frac{c}{\sin C}=d,$$

where a, b and c are the lengths of the sides of a triangle, and A, B and C are the opposite angles, while d is the diameter of the triangle's circumcircle. Also, the law is sometimes stated using the reciprocals

$$\frac{\sin A}{a}=\frac{\sin B}{b}=\frac{\sin C}{c}=\frac{1}{2R},$$

where R is the radius of the circumcircle.

■ **Example 5.32** Given the length of the sides $a=20$, $c=24$, and angle $C=40^\circ$, find the angle A. Using the law of sines, we conclude that

$$\frac{\sin A}{20}=\frac{\sin 40^\circ}{24}.$$

Then $A=\sin^{-1}\left(\frac{20\sin 40^\circ}{24}\right)\approx 32.39^\circ$.

■

■ **Example 5.33** If $\sin^2 B+\sin^2 C=\sin^2 A$, prove that the triangle ABC is right-angled.

Since we intend to apply Pythagorean identity, we use the law of sines to transform the given identity involving angles in an identity involving lengths. We obtain

$$\sin^2 B+\sin^2 C=\sin^2 A,$$

$$\left(\frac{b}{d}\right)^2+\left(\frac{c}{d}\right)^2=\left(\frac{a}{d}\right)^2, \qquad \text{(law of sines)}$$

$$b^2+c^2=a^2.$$

Consequently, the triangle ABC is right-angled.

■

5.5 The area of a triangle

In order to express the area of a triangle ABC in terms when two sides of a triangle are known, together with the angle between them, we often appeal to the formulae

$$Area=\frac{1}{2}bc\sin A=\frac{1}{2}ac\sin B=\frac{1}{2}ab\sin C,$$

Two other formulae for area are useful for the case in which the lengths of all sides are known instead.

(1) $Area=\sqrt{s(s-a)(s-b)(s-c)}$, in which $s=\frac{1}{2}(a+b+c)$.

(2) $Area=\frac{1}{4}\sqrt{2b^2c^2+2c^2a^2+2a^2b^2-a^4-b^4-c^4}$.

Exercise 5.8 In triangle ABC, if $a=3$, $b=4$ and $c=5$, then find the area of the triangle.

■

Solution Since $s=\frac{a+b+c}{2}=\frac{3+4+5}{2}=6$,

$$Area=\sqrt{s(s-a)(s-b)(s-c)}=\sqrt{6(6-3)(6-4)(6-5)}=\sqrt{6\cdot 3\cdot 2\cdot 1}=6$$

or

$$Area=\frac{1}{4}\sqrt{2b^2c^2+2c^2a^2+2a^2b^2-a^4-b^4-c^4}=\frac{1}{4}\sqrt{576}=6.$$

■

Exercise 5.9 Prove that, in any triangle ABC, $Area=2R^2\sin A\sin B\sin C$, where R is the radius of the circumcircle.

■

Proof

$$\begin{aligned} Area &=\frac{1}{2}bc\sin A \\ &=\frac{1}{2}\cdot 2R\sin B\cdot 2R\sin C\cdot \sin A \qquad \text{(law of sines)} \\ &=2R^2\sin A\sin B\sin C. \end{aligned}$$

■

5.6 The factor formulae

The set of 4 trigonometric identities below is known as the **factor formulae.** They convert expressions like $\sin A+\sin B$ into a product: similar to factorising the expression. Another common name for these identities is **sum and product formulae.**

$$\sin A+\sin B=2\sin\left(\frac{A+B}{2}\right)\cos\left(\frac{A-B}{2}\right);$$

$$\sin A-\sin B=2\cos\left(\frac{A+B}{2}\right)\sin\left(\frac{A-B}{2}\right);$$

$$\cos A+\cos B=2\cos\left(\frac{A+B}{2}\right)\cos\left(\frac{B-A}{2}\right);$$

$$\cos A-\cos B=2\sin\left(\frac{A+B}{2}\right)\sin\left(\frac{B-A}{2}\right).$$

Ⓡ Note carefully the symmetries and patterns in the four factor formulae. They can often be more easily remembered if you say the words as well as write the symbols, for example:

"sine plus sine equals twice the sine of half the sum, cosine of half the difference."

Ⓡ Note again the mnemonics: sin preserves the sign (hence the sign under the sin function in the right-hand side is the same as the sign in the left-hand side) and mixes the functions (hence the right-hand sides contain both a sin and a cos). Conversely, cos preserves the functions (it does not mix them, hence the coscos and sinsin terms in the right-hand side), but mixes the sign (hence the $\frac{B-A}{2}$ instead of $\frac{A-B}{2}$ in the right-hand side).

The factor formulae provide a powerful mathematical tool for dealing with trigonometric functions. They can be used to solve equations, simplify expressions and prove more identities.

■ **Example 5.34** Find $\sin 105°-\sin 15°$, without using a calculator.

$$\begin{aligned}\sin 105°-\sin 15°&=2\cos\left(\frac{105°+15°}{2}\right)\sin\left(\frac{105°-15°}{2}\right)\\&=2\cos 60°\sin 45°\\&=2\cdot\frac{1}{2}\cdot\frac{\sqrt{2}}{2}\\&=\frac{\sqrt{2}}{2}.\end{aligned}$$

■

Exercise 5.10 Prove that

$$\cos 2\theta+\cos 3\theta+\cos 4\theta=\cos 3\theta(1+2\cos\theta).$$

■

Proof One notes that the L.H.S. has three terms, one of which is $\cos 3\theta$. This also appears on the R.H.S., so it would be sensible to apply one of the factor formulae on the other two terms. Since

$$\cos 2\theta+\cos 4\theta=2\cos 3\theta\cos \theta,$$

$$\cos 2\theta+\cos 3\theta+\cos 4\theta=2\cos 3\theta\cos \theta+\cos 3\theta=\cos 3\theta(1+2\cos \theta).$$

■

Exercise 5.11 Evaluate

$$2\cos 75^\circ \cos 15^\circ.$$

■

Solution Notice the two cosine terms. The factor formulae suggest that this is created form the sum of two cosines. If $2\cos 75^\circ \cos 15^\circ=\cos x+\cos y$, then

$$x+y=150^\circ \text{ and } x-y=30^\circ,$$

which implies that

$$2\cos 75^\circ\cos 15^\circ=\cos 90^\circ+\cos 60^\circ=\frac{1}{2}.$$

■

Exercise 5.12 Prove that, in any triangle ABC, $b\cos B+c\cos C=a\cos (B-C)$. ■

Proof

$$\begin{aligned}
b\cos B+c\cos C &=d\sin B\cos B+d\sin C\cos C && \text{(law of sines)}\\
&=\frac{d}{2}(2\sin B\cos B+2\sin C\cos C)\\
&=\frac{d}{2}(\sin 2B+\sin 2C) && \text{(double angle formula)}\\
&=\frac{d}{2}2\sin (B+C)\cos (B-C) && \text{(factor formula)}\\
&=d\sin (\pi-A)\cos (B-C)\\
&=d\sin A\cos (B-C) && \text{(supplementary angle formula)}
\end{aligned}$$

Similarly,

$$a\cos (B-C)=d\sin A \cos (B-C), \qquad \text{(law of sines)}$$

from which the conclusion follows.

■

5.7 General solutions of trigonometric equations

Definition 5.7.1—Trigonometric equation. A trigonometric equation is an equation involving one or more trigonometric functions of a variable.

By solving a trigonometric equation, we mean to find the angle or angles which satisfy the equation. The values of the variable (angle) are known as the (roots) of the equation.

The general solution of $\sin x=k(-1\leqslant k\leqslant 1)$

Let θ be a particular angle whose sine is k. Then, $\pi-\theta$ is the other angle having the same sine. Since the value of the trigonometric function is unaltered by adding or subtracting an integral multiple of 2π to or from θ, we have

$$x=2m\pi+\theta=2m\pi+(-1)^{2m}\theta$$

and

$$x=2m\pi+\pi-\theta=(2m+1)\pi+(-1)^{2m+1}\theta,$$

in which m is an integer, as the solutions of the equation

$$\sin x=k.$$

Combining the two equations above , we have

$$x=n\pi+(-1)^n\theta$$

for any integer n.

The general solution of $\cos x=k(-1\leqslant k\leqslant 1)$

Let θ be a particular angle such that $\cos\theta=k$. Then $-\theta$ is another value of x whose cosine is the the same constant k. Since all coterminal angles of θ and $-\theta$ are given by $2n\pi+\theta$ and $2n\pi-\theta$, $\cos x=\cos(2n\pi\pm\theta)$. Hence, the general solution is given by

$$x=2n\pi\pm\theta \text{ for any integer } n.$$

The general solution of $\tan\theta=k$

Let θ be a particular angle so that $\tan\theta=k$. The other values of x for which $\tan x=k$ is $\pi+\theta$. Thus the general solution is given by

$$x=n\pi+\theta \text{ for any integer } n.$$

Exercise 5.13 Solve the equation $2\cos^2 x-5\cos x+2=0$ for $0\leqslant x\leqslant 2\pi$. ■

Solution Factoring the left hand side, we get

$$(\cos x-2)(2\cos x-1)=0.$$

Thus

$$\cos x-2=0 \quad\text{and}\quad 2\cos x-1=0$$

or

$$\cos x=2 \quad\text{and}\quad \cos x=\frac{1}{2}.$$

Since $\cos x$ can never be greater than (numerically) 1, $\cos x=2$ has no solution. The equation $\cos x=\frac{1}{2}$ has $\theta=\frac{\pi}{3}$ as a particular solution and hence $x=\pm\frac{\pi}{3}+2n\pi$, for any integer n, is its general solution. The solutions in the given range are

$$x_1=\frac{\pi}{3} \text{ and } x_2=-\frac{\pi}{3}+2\pi=\frac{5\pi}{3}.$$

■

Exercise 5.14 Solve the equation $\cot x+\tan x=2$.

■

Solution The given equation is same as

$$\frac{1}{\tan x}+\tan x=2,$$

which implies that

$$(\tan x-1)^2=0.$$

Since $\tan\frac{\pi}{4}=1$, $x=n\pi+\frac{\pi}{4}$ for any integer n is the required general solution.

■

Exercise 5.15 Solve the equation $\cos 3x+\cos x=\cos 2x$.

■

Solution Applying one of the factor formulae on $\cos 3x+\cos x$, we get

$$2\cos 2x\cos x-\cos 2x=0,$$

that is,

$$\cos 2x(2\cos x-1)=0.$$

Then either $\cos 2x=0$ or $2\cos x-1=0$. Hence, for any integer n,

$$x=(2n+1)\frac{\pi}{4} \text{ or } x=2n\pi\pm\frac{\pi}{3}.$$

■

Exercise 5.16 Solve the equation $\sin^2\theta-2\cos\theta+\frac{1}{4}=0$.

■

Solution Since $\sin^2\theta=1-\cos^2\theta$, we obtain

$$4-4\cos^2\theta-8\cos\theta+1=0,$$

which implies that

$$(2\cos\theta+5)(2\cos\theta-1)=0.$$

Then either $2\cos\theta+5=0$ or $2\cos\theta-1=0$. Since $\cos\theta\geqslant-1$, $\cos\theta=-\frac{5}{2}$ is not possible. The equation $2\cos\theta-1=0$ has $\theta=\frac{\pi}{3}$ as a particular solution. Its general solution is then

$$\theta=2n\pi\pm\frac{\pi}{3}$$

for any integer n.

■

5.8 Summary

Angle measurement

$$\pi=180^\circ;$$

$$1^\circ=\frac{\pi}{180};$$

$$1\text{ rad}=\frac{180}{\pi}.$$

Right triangle trigonometry

$$\sin A=\frac{a}{h}=\frac{\text{opposite}}{\text{hypotenuse}};$$

$$\cos A=\frac{b}{h}=\frac{\text{adjacent}}{\text{hypotenuse}};$$

$$\tan A=\frac{a}{b}=\frac{\text{opposite}}{\text{adjacent}};$$

$$\cot A=\frac{b}{a}=\frac{\text{adjacent}}{\text{opposite}}=\frac{1}{\tan A};$$

$$\sec A=\frac{h}{b}=\frac{\text{hypotenuse}}{\text{adjacent}}=\frac{1}{\cos A};$$

$$\csc A=\frac{h}{a}=\frac{\text{hypotenuse}}{\text{opposite}}=\frac{1}{\sin A}.$$

Trigonometric functions

$$\sin\theta=\frac{y}{r},\ \csc\theta=\frac{r}{y};$$

$$\cos\theta=\frac{x}{y},\ \sec\theta=\frac{r}{x};$$

$$\tan\theta=\frac{y}{x},\ \cot\theta=\frac{x}{y}.$$

The periodicity of the sine and cosine functions

$\sin\theta=\sin(\theta+2k\pi)$, for any angle θ and any inteser k;

$\cos\theta=\cos(\theta+2k\pi)$, for any angle θ and any integer k.

Maximal and minimal values of the sine and cosine functions

$-1\leqslant\sin\theta\leqslant 1$, for any angle θ;

$-1\leqslant\cos\theta\leqslant 1$, for any angle θ.

Values of the sine and cosine functions for common angles

$\sin k\pi=0$, for any integer k;

$$\cos\frac{(2k+1)\pi}{2}=0, \text{ for any integer } k.$$

That is, the sine of any integer multiple of π and the cosine of any odd multiple of $\frac{\pi}{2}$ are 0.

x	$0°$	$30°$	$45°$	$60°$	$90°$
$\sin x$	$\frac{\sqrt{0}}{2}$	$\frac{\sqrt{1}}{2}$	$\frac{\sqrt{2}}{2}$	$\frac{\sqrt{3}}{2}$	$\frac{\sqrt{4}}{2}$
$\cos x$	$\frac{\sqrt{4}}{2}$	$\frac{\sqrt{3}}{2}$	$\frac{\sqrt{2}}{2}$	$\frac{\sqrt{1}}{2}$	$\frac{\sqrt{0}}{2}$

The signs of the sine and cosine functions

$$\sin\theta\geqslant 0,\ \theta\in[0,\pi];$$
$$\sin\theta\leqslant 0,\ \theta\in[\pi, 2\pi];$$
$$\cos\theta\geqslant 0,\ \theta\in\left[0, \frac{\pi}{2}\right]\cup\left[\frac{3\pi}{2}, 2\pi\right];$$
$$\cos\theta\leqslant 0,\ \theta\in\left[\frac{\pi}{2},\frac{3\pi}{2}\right].$$

Pythagorean trigonometric identities

$$\sin^2\theta+\cos^2\theta=1;$$
$$\sec^2\theta=1+\tan^2\theta\Rightarrow 1+\tan^2\theta=\frac{1}{\cos^2\theta};$$
$$\csc^2\theta=1+\cot^2\theta\Rightarrow 1+\cot^2\theta=\frac{1}{\sin^2\theta}.$$

Addition and subtraction formulae for the sine and cosine functions

$$\sin(A+B)=\sin A\cos B+\cos A\sin B;$$
$$\cos(A+B)=\cos A\cos B-\sin A\sin B;$$
$$\sin(A-B)=\sin A\cos B-\cos A\sin B;$$
$$\cos(A-B)=\cos A\cos B+\sin A\sin B.$$

Addition and subtraction formulae for the tangent function

$$\tan(A+B)=\frac{\tan A+\tan B}{1-\tan A\tan B};$$
$$\tan(A-B)=\frac{\tan A-\tan B}{1+\tan A\tan B}.$$

Parity of the sine, cosine and tangent function

$\sin(-B)=-\sin B$ (the sine function is an odd function);

$\cos(-B)=\cos B$ (the cosine function is an even function);

$\tan(-B)=-\tan B$ (the tangent function is an odd function).

Functions of the complementary angle

$$\sin\left(\frac{\pi}{2}-B\right)=\cos B;$$

$$\cos\left(\frac{\pi}{2}-B\right)=\sin B;$$

$$\tan\left(\frac{\pi}{2}-B\right)=\cot B.$$

Functions of the supplementary angle

$$\sin(\pi-B)=\sin B;$$
$$\cos(\pi-B)=-\cos B;$$
$$\tan(\pi-B)=-\tan B.$$

$$\sin(\pi+B)=-\sin B;$$
$$\cos(\pi+B)=-\cos B;$$
$$\tan(\pi+B)=\tan B.$$

Double angle formulae

$$\sin 2A=2\sin A\cos A;$$
$$\begin{aligned}\cos 2A&=\cos^2 A-\sin^2 A\\&=2\cos^2 A-1\\&=1-2\sin^2 A;\end{aligned}$$
$$\tan 2A=\frac{2\tan A}{1-\tan^2 A}.$$
$$\cos^2 A=\frac{1+\cos 2A}{2}$$
$$\sin^2 A=\frac{1-\cos 2A}{2}.$$

Half angle formulae

$$\tan A=\frac{2\tan\frac{A}{2}}{1-\tan^2\frac{A}{2}};$$

$$\sin A=\frac{2\tan\frac{A}{2}}{1+\tan^2\frac{A}{2}};$$

$$\cos A=\frac{1-\tan^2\frac{A}{2}}{1+\tan^2\frac{A}{2}}.$$

Triple-angle formulae

$$\sin 3A=3\sin A-4\sin^3 A;$$

$$\cos 3A=4\cos^3 A-3\cos A;$$

$$\tan 3A=\frac{3\tan A-\tan^3 A}{1-3\tan^2 A}.$$

Law of cosines

If the angles α, β and γ are respectively opposite the sides a, b and c, then

$$\cos\alpha=\frac{b^2+c^2-a^2}{2bc} \text{ or } a^2=b^2+c^2-2bc\cos\alpha;$$

$$\cos\beta=\frac{a^2+c^2-b^2}{2ac} \text{ or } b^2=a^2+c^2-2ac\cos\beta;$$

$$\cos\gamma=\frac{a^2+b^2-c^2}{2ab} \text{ or } c^2=a^2+b^2-2ab\cos\gamma.$$

Law of sines

$$\frac{a}{\sin A}=\frac{b}{\sin B}=\frac{c}{\sin C}=d,$$

where d is the diameter of the triangle's circumcircle.

The area of a triangle

$$Area=\frac{1}{2}bc\sin A=ac\sin B=ab\sin C.$$

(1) $Area=\sqrt{s(s-a)(s-b)(s-c)}$, in which $s=\frac{1}{2}(a+b+c)$.

(2) $Area=\frac{1}{4}\sqrt{2b^2c^2+2c^2a^2+2a^2b^2-a^4-b^4-c^4}$.

The factor formulae

$$\sin A+\sin B=2\sin\left(\frac{A+B}{2}\right)\cos\left(\frac{A-B}{2}\right);$$

$$\sin A-\sin B=2\cos\left(\frac{A+B}{2}\right)\sin\left(\frac{A-B}{2}\right);$$

$$\cos A+\cos B=2\cos\left(\frac{A+B}{2}\right)\cos\left(\frac{B-A}{2}\right);$$

$$\cos A-\cos B=2\sin\left(\frac{A+B}{2}\right)\sin\left(\frac{B-A}{2}\right).$$

The general solution of $\sin x=k(-1\leqslant k\leqslant 1)$

Let θ be a particular angle whose sine is k. The general solution is given by

$$x=n\pi+(-1)^n\theta \text{ for any integer } n.$$

The general solution of $\cos x=k(-1\leqslant k\leqslant 1)$

Let θ be a particular angle such that $\cos\theta=k$. The general solution is given by

$$x=2n\pi\pm\theta \text{ for any integer } n.$$

The general solution of $\tan\theta=k$

Let θ be a particular angle so that $\tan\theta=k$. The general solution is given by

$$x=n\pi+\theta \text{ for any integer } n.$$

5.9 Further problems

Problem 5.1 Given that $\tan\theta=\frac{3}{4}$, find the value of $\frac{\sin\theta}{\cos\theta+\sin\theta}$.

Problem 5.2 Prove the identities:

(1) $\sin\theta\tan\theta=\sec\theta-\cos\theta$;

(2) $\frac{\cos\theta}{1-\sin\theta}=\frac{1+\sin\theta}{\cos\theta}$;

(3) $\frac{1}{\sec\theta-\tan\theta}=\sec\theta+\tan\theta$.

Problem 5.3 Evaluate, without using tables,

(1) $\sin\theta$ and $\cos\theta$, if $\tan\theta=-\frac{3}{4}$;

(2) $\tan\theta$ and $\sin\theta$, if $\cos\theta=-\frac{4}{5}$;

(3) $\cos\theta$ and $\tan\theta$, if $\sin\theta=-\frac{24}{25}$;

(4) $\cos\theta$ and $\tan\theta$, if $\sin\theta=\frac{5}{13}$.

Problem 5.4 Find $\sin A$ and $\cos A$ if $\tan A=\frac{3}{4}$ and A is acute.

Problem 5.5 Find $\sin A$ and $\tan A$ if $\cos A=-\frac{4}{5}$ and $180°<A<270°$.

Problem 5.6 Evaluate $\sin\left(\frac{\pi}{6}+\frac{\pi}{4}\right)+\sin\left(\frac{\pi}{6}-\frac{\pi}{4}\right)$.

Problem 5.7 Using compound angles, find the values of the following in surd form,

(1) $\sin 15°$; (2) $\cos 15°$; (3) $\sin 105°$;

(4) $\tan 105°$; (5) $\sin\frac{5\pi}{6}$.

Problem 5.8 Prove that $\tan\left(\frac{\pi}{4}+\theta\right)=\frac{\cos\theta+\sin\theta}{\cos\theta-\sin\theta}$.

Problem 5.9 Prove that $\sin 15°\cos 15°=\frac{1}{4}$.

Problem 5.10 Find the exact value of $\frac{2\tan 67.5°}{1-\tan^2 67.5°}$.

Problem 5.11 If $\sin\alpha+\cos\alpha=\frac{1}{3}$, find $\sin 2\alpha$.

Problem 5.12 Find $\cos 2\theta$ knowing that $\sin\theta=\frac{1}{3}$.

Problem 5.13 Evaluate $\sin 75°+\sin 15°$.

Problem 5.14 Solve the triangle ABC given $B=79°$, $AC=22.31$ mm and $AB=17.92$ mm. Find also its area.

Problem 5.15 The sides of a triangle are 3, 7 and 8 units respectively. Prove that one of the angles is 60°and calculate the other two angles to the nearest degree.

Problem 5.16 In a triangle ABC, if $a=12$ cm, $b=20$ cm and $c=14$ cm, find A, B and C.

Problem 5.17 In a triangle ABC, if $A=35°$, $a=17$ cm and $B=50°$, find b.

Problem 5.18 In a triangle ABC, if $A=60°$, $a=6$ cm and $b=5$ cm, find C.

Problem 5.19 Solve the following equations:

(1) $4\sin\theta-3\cos\theta=0$ for $\theta\in[0,2\pi]$;

(2) $\cos\theta\sin\theta=2\sin\theta$ for $\theta\in[0,2\pi]$.

Problem 5.20 Solve the following equations:

(1) $\cos(\theta+30°)=\frac{1}{2}$ for $\theta\in[-\pi,\pi]$;

(2) $\tan 2\theta=1$ for $\theta\in[0,2\pi]$.

Problem 5.21 Solve the equation $\sin(\theta+45°)=3\cos(\theta+45°)$ for $0°\leqslant\theta\leqslant 360°$.

Answers

Problem 5.1 Divide by $\cos\theta$ both the numerator and denominator. The value is $\frac{3}{7}$.

Problem 5.2 (1) $\sin\theta\tan\theta=\frac{\sin^2\theta}{\cos\theta}=\frac{1-\cos^2\theta}{\cos\theta}=\sec\theta-\cos\theta$.

(2) $\frac{\cos\theta}{1-\sin\theta}=\frac{\cos\theta(1+\sin\theta)}{\cos^2\theta}=\frac{1+\sin\theta}{\cos\theta}$.

(3) $\sec^2\theta-\tan^2\theta=(\sec\theta-\tan\theta)(\sec\theta+\tan\theta)=1$.

Problem 5.3 (1) $-\frac{3}{5}$, $\frac{4}{5}$ or $\frac{3}{5}$, $-\frac{4}{5}$.

(2) $\frac{3}{4}$, $-\frac{3}{5}$ or $-\frac{3}{4}$, $\frac{3}{5}$.

(3) $\frac{7}{25}$, $-\frac{24}{7}$ or $-\frac{7}{25}$, $\frac{24}{7}$.

(4) $\frac{12}{13}$, $\frac{5}{12}$ or $-\frac{12}{13}$, $-\frac{5}{12}$.

Problem 5.4 $\frac{3}{5}$, $\frac{4}{5}$.

Problem 5.5 $-\frac{3}{5}$, $\frac{3}{4}$.

Problem 5.6 $\frac{\sqrt{2}}{2}$.

Problem 5.7 (1) $\sin(45^\circ-30^\circ)=\frac{\sqrt{3}-1}{2\sqrt{2}}$.

(2) $\cos(45^\circ-30^\circ)=\frac{\sqrt{3}+1}{2\sqrt{2}}$.

(3) $\sin(60^\circ+45^\circ)=\frac{\sqrt{3}+1}{2\sqrt{2}}$.

(4) $\tan(60^\circ+45^\circ)=-2-\sqrt{3}$.

(5) $\sin\left(\pi-\frac{\pi}{6}\right)=\frac{1}{2}$.

Problem 5.8 Use the addition formula for the tangent function and multiply both the numerator and denominator by $\cos\theta$.

Problem 5.9 $\sin 15^\circ\cos 15^\circ=\frac{1}{2}2\sin 15^\circ\cos 15^\circ=\frac{1}{2}\sin 30^\circ$.

Problem 5.10 $\tan 135^\circ=-1$.

Problem 5.11 Raise to the power of 2 and use the double angle formula and the fundamental Pythagorean trigonometric identity; $-\frac{8}{9}$.

Problem 5.12 Use $\cos 2\theta=1-2\sin^2\theta$; $\frac{7}{9}$.

Problem 5.13 Use factor formula; $\frac{\sqrt{6}}{2}$.

Problem 5.14 $A=50$, $C=51°$, $BC=17.41$ mm, $Area=153.12$ mm^2.

Problem 5.15 22°, 98°.

Problem 5.16 $A=36.2°$, $B=100.3°$ and $C=43.5°$.

Problem 5.17 22.7 cm.

Problem 5.18 73.8°.

Problem 5.19 (1) 36.9°. (2) $0, \pi, 2\pi$.

Problem 5.20 (1) $-90°$, 30°. (2) 22.5°, 112.5°, 202.5°, 292.5°.

Problem 5.21 $\tan\theta=\frac{1}{2}\Rightarrow\theta=18.4°$, 198.4°.

Chapter 6 Vectors

6.1 Vector geometry

Definition 6.1.1—Scalar. In mathematics, a **scalar** refers to a quantity described by a single real number, used to measured magnitude (size).

Examples of quantities described by scalars are mass, voltage and temperature, and so on.

Definition 6.1.2—Vector. A quantity which requires both magnitude and direction to be described completely is called a **vector**.

Examples of quantities described by vectors are velocity, force and acceleration.

Magnitude and direction

A vector is frequently represented by a directed line segment (the order of its endpoints matters, that is), connecting an **initial point** (or **tail**) with a **terminal point** (or **head**). The vector with initial point A and terminal point B is denoted by $\overrightarrow{AB}$. The **magnitude** (also called **length** or **Euclidean norm**) of $\overrightarrow{AB}$, denoted by $\| \overrightarrow{AB} \|$, is defined to be the length of the segment AB. The **direction** of $\overrightarrow{AB}$ is defined to be the direction of the line segment AB.

Equality of vectors

As shown in Figure 6.1, two vectors $\overrightarrow{AB}$ and $\overrightarrow{CD}$ are said to be **equal** or **equivalent** if they have the same magnitude (that is, $\| \overrightarrow{AB} \| = \| \overrightarrow{CD} \|$) and direction (that is, the support lines AB and CD are either parallel or coincide and the directions from A to B and from C to D provide the same orientation on those lines).

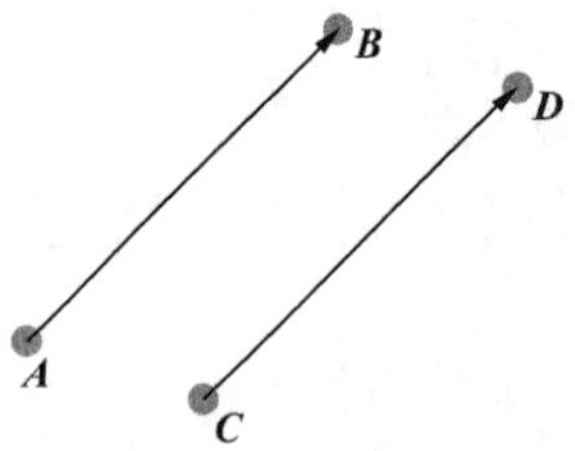

Figure 6.1 Two equal vectors $\overrightarrow{AB}$ and $\overrightarrow{CD}$.

The null vector

If A coincides with B, one obtains the **null vector**, or **zero vector**, also denoted as $\mathbf{0}$, which has length 0 and whose direction is undefined.

Unit vector

A vector of length 1 is called an **unit vector.** Dividing any nonzero vector by its length, one obtains a unit vector.

Notation 6.1 If its initial and terminal point are not specified, a vector is usually denoted in lowercase boldface, as in $\boldsymbol{a}$, $\boldsymbol{b}$, $\boldsymbol{c}$, $\boldsymbol{u}$, $\boldsymbol{v}$, or with non-bold lowercase accented with a small arrow, as in $\vec{a}$, $\vec{b}$, $\vec{c}$, $\vec{u}$, $\vec{v}$.

The negative vector

Given a vector $\boldsymbol{a}$, a vector which has the same length as $\boldsymbol{a}$, but opposite direction, is denoted by $-\boldsymbol{a}$. For instance, as shown in Figure 6.2, the negative of a vector $\overrightarrow{AB}$ is the vector $\overrightarrow{BA}$. That is, $\overrightarrow{BA}=-\overrightarrow{AB}$.

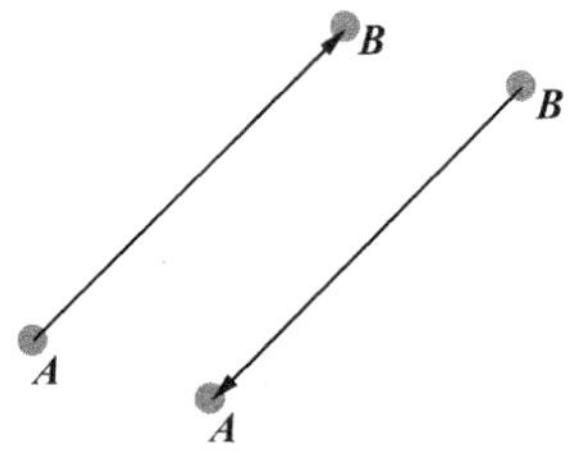

Figure 6.2 $\overrightarrow{BA}=-\overrightarrow{AB}$.

Ⓡ Remember that, for a vector, the order of endpoints matters. Changing the order the endpoint multiplies the vector by -1, as seen above in the formula $\overrightarrow{BA}=-\overrightarrow{AB}$.

Multiplication of a vector by a scalar

If c is a real number and $\boldsymbol{v}$ is a vector, then $c\boldsymbol{v}$ is a vector of magnitude $|c|\|\boldsymbol{v}\|$ which has the same direction as $\boldsymbol{v}$ if $c>0$ and opposite direction if $c<0$. An example is illustrated in Figure 6.3.

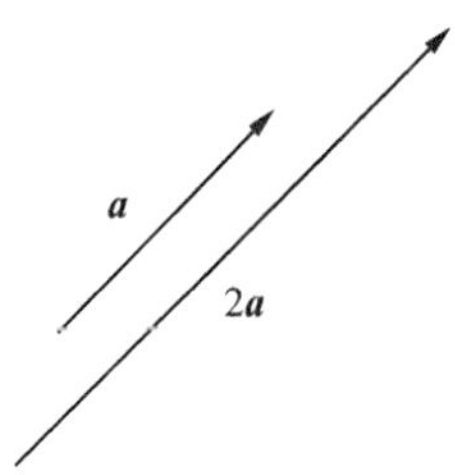

Figure 6.3 Scalar multiplication.

Parallel (collinear) vectors

Two vectors $\boldsymbol{a}$ and $\boldsymbol{b}$ are parallel or collinear if they have the same direction. In this case, one of them is obtained from the other via multiplication with a constant.

Addition of vectors

Triangle law

To add vectors, one can use the **triangle law** (or **head-to-tail law**). Given two vectors $\boldsymbol{a}$ and $\boldsymbol{b}$, their sum $\boldsymbol{a}+\boldsymbol{b}$ may be represented graphically, as seen in Figure 6.4 by placing the tail of the vector $\boldsymbol{b}$ at the head of the vector $\boldsymbol{a}$, and then drawing a vector from the tail of $\boldsymbol{a}$ to the head of $\boldsymbol{b}$. The new vector drawn represents the vector $\boldsymbol{a}+\boldsymbol{b}$. This vector closes a triangle, hence the name of the rule.

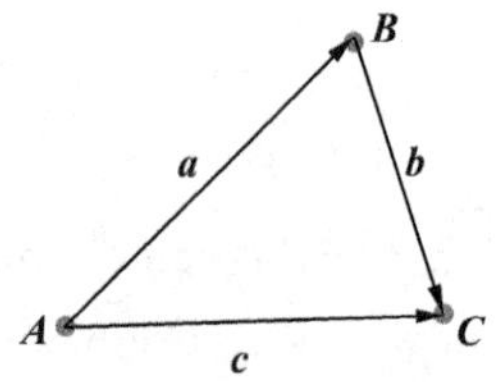

Figure 6.4 The sum of *a* and *b* by triangle law.

® One does not always need diagrams if letter correspondence is observed. For instance,

$$\overrightarrow{AB}+\overrightarrow{BC}=\overrightarrow{AC},$$

since the tail of the second vector, B, coincides with the head of the first vector. Then the sum vector, $\overrightarrow{AC}$, starts in the first point, A, and ends in the last point, C.

® Notice the mnemonics: for vectors, the straight way ($\overrightarrow{AC}$, directly from A to C) equals the deviating way ($\overrightarrow{AB}+\overrightarrow{BC}$, deviating through B).

Parallelogram law

To add vectors, one may also use the **parallelogram law**. Given two vectors $\boldsymbol{a}$ and $\boldsymbol{b}$, place the tail of $\boldsymbol{b}$ at the tail of $\boldsymbol{a}$ and draw the parallelogram determined by the sides $\boldsymbol{a}$ and $\boldsymbol{b}$. Then, as shown in Figure 6.5, the sum $\boldsymbol{a}+\boldsymbol{b}$ is the diagonal of the parallelogram starting from the common tail.

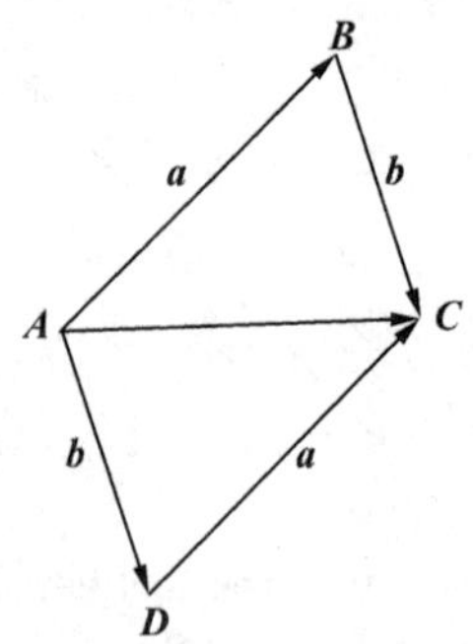

Figure 6.5 The sum of *a* and *b* by parallelogram law.

Polygon law

To add more than two vectors, one can use the **polygon law**, which consists in the multiple use of the **triangle law**, by placing repeatedly the tail of the next vector at the head of the previous vector and then drawing a vector from the tail of the first vector to the head of the last vector. The newly drawn vector, which closes the polygon, represents the sum of all previous vectors (see Figure 6.6).

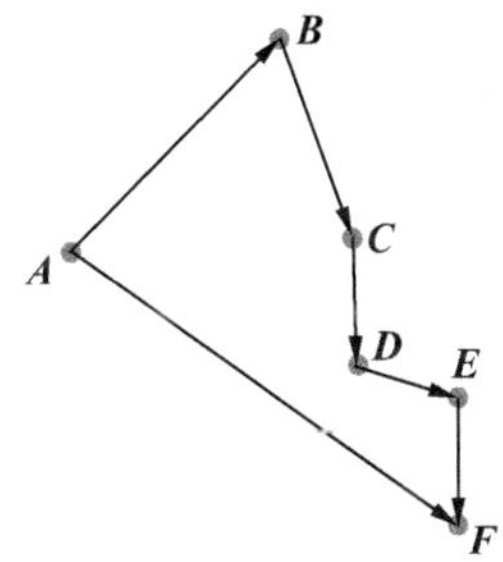

Figure 6.6 The sum of multiple vectors by polygon law.

® Again, one does not always need diagrams if letter correspondence is observed. For instance,

$$\overrightarrow{AB}+\overrightarrow{BC}+\overrightarrow{CD}=\overrightarrow{AD}$$

by the repeated application of the triangle rule.

® Notice again the mnemonics: the straight way ($\overrightarrow{AD}$, directly from A to D) equals the deviating way ($\overrightarrow{AB}+\overrightarrow{BC}+\overrightarrow{CD}$, deviating through both B and C). One may deviate through as many points as necessary.

® If one adds together a set of vectors which return to the starting point, while the tail of the next vector is always at the head of the previous vector, then the result is the null vector. For instance, $\overrightarrow{AB}+\overrightarrow{BC}+\overrightarrow{CA}=0$, as the vectors start from A and return also to A.

Subtraction of vectors

To find $\boldsymbol{a}-\boldsymbol{b}$, one simply adds $\boldsymbol{a}$ with the negative of $\boldsymbol{b}$. If $\boldsymbol{a}$ and $\boldsymbol{b}$ have the same tail, then $\boldsymbol{a}-\boldsymbol{b}$ is the vector drawn from the head of $\boldsymbol{b}$ to the head of $\boldsymbol{a}$ (see Figure 6.7).

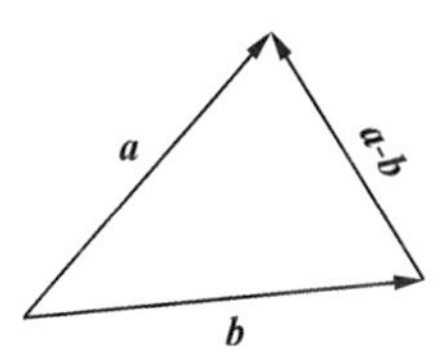

Figure 6.7 The difference $a-b$.

® Notice the mnemonics: "departs from negative, arrives to positive" ($\boldsymbol{a}-\boldsymbol{b}$ departs from the head of $\boldsymbol{b}$ and arrives at the head of $\boldsymbol{a}$), or "arrival minus departure" (the vector which arrives at the head of $\boldsymbol{a}$ and departs from the head of $\boldsymbol{b}$ is $\boldsymbol{a}-\boldsymbol{b}$).

6.2 Vector algebra

Although we have introduced the notion of a vector together with ways to perform operations between vectors using geometric considerations, it is possible to operate from an entirely algebraic viewpoint instead. To this purpose, the standard basis provides a convenient way to represent all vector using numbers alone.

6.2.1 Vectors in three dimensions

Definition 6.2.1—Standard basis. We shall call the **standard basis** (**natural basis**) the set of unit vectors $\{\boldsymbol{i}, \boldsymbol{j}, \boldsymbol{k}\}$ which have the direction of the axes of a Cartesian coordinate system, as shown in Figure 6.8.

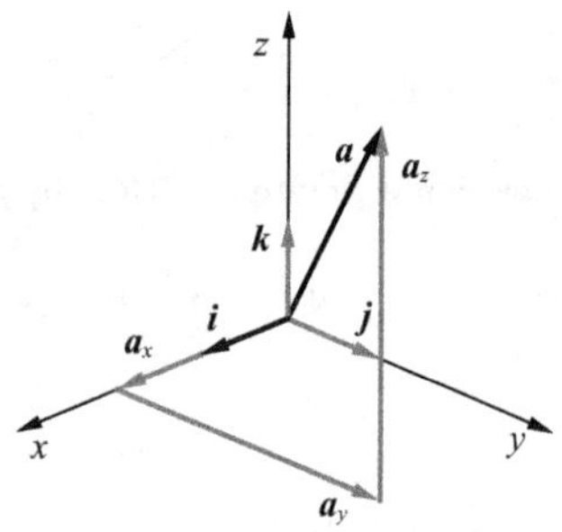

Figure 6.8 Standard basis vectors.

Ⓡ Other notations for the standard basis, such as $\{\boldsymbol{e}_x, \boldsymbol{e}_y, \boldsymbol{e}_z\}$, $\{\boldsymbol{e}_1, \boldsymbol{e}_2, \boldsymbol{e}_3\}$, $\{\vec{i}, \vec{j}, \vec{k}\}$ are also in use.

We are now ready to characterize vectors using sets of numbers.

The components of a vector

For any vector $\boldsymbol{a}$, there exist the numbers a_1, a_2, a_3, which are uniquely determined, such that

$$\boldsymbol{a}=a_1\boldsymbol{i}+a_2\boldsymbol{j}+a_2\boldsymbol{k}.$$

The numbers a_1, a_2, a_3 are called the **components** (or **coordinates**) of $\boldsymbol{a}$ with respect to the standard basis and, in this context, the vector $\boldsymbol{a}$ may also represented by indicating its components, in the form

$$\boldsymbol{a}=(a_1, a_2, a_3).$$

The vector $\boldsymbol{a}$ is completely determined giving its components.

■ **Example 6.1** The components of $\boldsymbol{a}=6\boldsymbol{i}-3\boldsymbol{j}+2\boldsymbol{k}$ are 6, −3, 2. The vector $\boldsymbol{a}$ may also be represented as $\boldsymbol{a}=(6, 3, 2)$.

■

Equality of two vectors

Two vectors $\boldsymbol{a}$ and $\boldsymbol{b}$ are equal if and only if they have the same components.

■ **Example 6.2** If $\boldsymbol{a}=(3x+1, 2y+3, z+4)$ and $\boldsymbol{a}=(-8, 5, 2)$, find x, y and z so that $\boldsymbol{a}=\boldsymbol{b}$.

We use the fact that the vectors $\boldsymbol{a}$ and $\boldsymbol{b}$ have equal components.

$$\boldsymbol{a}=\boldsymbol{b} \Rightarrow (3x+1, 2y+3, z+4)=(-8, 5, 2).$$

Consequently,

$$3x+1=-8 \Rightarrow x=-3,$$
$$2y+3=5 \Rightarrow y=1,$$
$$z+4=2 \Rightarrow z=-2.$$

■

The magnitude of a vector

If $\boldsymbol{a}=a_1\boldsymbol{i}+a_2\boldsymbol{j}+a_3\boldsymbol{k}$, then the magnitude of $\boldsymbol{a}$ is

$$\|\boldsymbol{a}\|=\sqrt{a_1^2+a_2^2+a_3^2}.$$

■ **Example 6.3** Find the magnitude of the vector $\boldsymbol{a}=4\boldsymbol{i}-2\boldsymbol{j}+4\boldsymbol{k}$.

The magnitude is

$$\|\boldsymbol{a}\|=\sqrt{4^2+(-2)^2+4^2}=\sqrt{36}=6.$$

■

The null vector

The null vector is a vector with zero components, that is

$$\mathbf{0}=0\boldsymbol{i}+0\boldsymbol{j}+0\boldsymbol{k} \quad \text{or} \quad \mathbf{0}=(0,0,0).$$

The null vector has zero magnitude.

The negative or inverse of a vector

If $\boldsymbol{a}=a_1\boldsymbol{i}+a_2\boldsymbol{j}+a_3\boldsymbol{k}=(a_1, a_2, a_3)$, then its negative is

$$-\boldsymbol{a}=-a_1\boldsymbol{i}-a_2\boldsymbol{j}-a_3\boldsymbol{k} \text{ or } -\boldsymbol{a}=(-a_1, -a_2, -a_3).$$

Addition and subtraction of vectors

The addition and subtraction of vectors can be performed using the respective operations on their components. If $\boldsymbol{a}=a_1\boldsymbol{i}+a_2\boldsymbol{j}+a_3\boldsymbol{k}$ and $\boldsymbol{b}=b_1\boldsymbol{i}+b_2\boldsymbol{j}+b_3\boldsymbol{k}$, then

$$\boldsymbol{a}+\boldsymbol{b}=(a_1+b_1)\boldsymbol{i}+(a_2+b_2)\boldsymbol{j}+(a_3+b_3)\boldsymbol{k};$$
$$\boldsymbol{a}-\boldsymbol{b}=(a_1-b_1)\boldsymbol{i}+(a_2-b_2)\boldsymbol{j}+(a_3-b_3)\boldsymbol{k}.$$

■ **Example 6.4** If $\boldsymbol{a}=3\boldsymbol{i}+\boldsymbol{j}+6\boldsymbol{k}$ and $\boldsymbol{b}=4\boldsymbol{i}-5\boldsymbol{j}-\boldsymbol{k}$, then

$$\boldsymbol{a}+\boldsymbol{b}=(3\boldsymbol{i}+\boldsymbol{j}+6\boldsymbol{k})+(4\boldsymbol{i}-5\boldsymbol{j}-\boldsymbol{k})=7\boldsymbol{i}-4\boldsymbol{j}+5\boldsymbol{k};$$
$$\boldsymbol{a}-\boldsymbol{b}=(3\boldsymbol{i}+\boldsymbol{j}+6\boldsymbol{k})-(4\boldsymbol{i}-5\boldsymbol{j}-\boldsymbol{k})=-\boldsymbol{i}+6\boldsymbol{j}+7\boldsymbol{k}.$$

■

Exercise 6.1 If A, B, C are such that $\overrightarrow{AB}=(3,4,-2)$ and $\overrightarrow{BC}=(-1,5,7)$, then find $\overrightarrow{CA}$.

■

Solution The straight way ($\overrightarrow{CA}$, directly from C to A) equals the deviating way ($\overrightarrow{CB}+\overrightarrow{BA}$, deviating through B). It follows that

$$\begin{aligned}\overrightarrow{CA}&=\overrightarrow{CB}+\overrightarrow{BA}\\&=-\overrightarrow{BC}-\overrightarrow{AB}\\&=(1,-5,-7)-(3,4,-2)=(-2,-9,-5).\end{aligned}$$

■

Scalar multiplication

The scalar multiplication of a vector can be computed multiplying each component with that constant. If $\boldsymbol{a}=a_1\boldsymbol{i}+a_2\boldsymbol{j}+a_3\boldsymbol{k}$ and c is a constant, then

$$c\boldsymbol{a}=ca_1\boldsymbol{i}+ca_2\boldsymbol{j}+ca_3\boldsymbol{k}.$$

■ **Example 6.5** If $\boldsymbol{a}=3\boldsymbol{i}+4\boldsymbol{j}+5\boldsymbol{k}$ and $\boldsymbol{b}=10\boldsymbol{i}-2\boldsymbol{j}$, then find $\boldsymbol{b}-2\boldsymbol{a}$.

We multiply the coordinates of $\boldsymbol{a}$ by 2 and subtract them from the coordinates of $\boldsymbol{a}$. One obtains

$$\begin{aligned}\boldsymbol{b}-2\boldsymbol{a}&=10\boldsymbol{i}-2\boldsymbol{j}-2(3\boldsymbol{i}+4\boldsymbol{j}+5\boldsymbol{k})\\&=10\boldsymbol{i}-2\boldsymbol{j}-(6\boldsymbol{i}+8\boldsymbol{j}+10\boldsymbol{k})\\&=4\boldsymbol{i}-10\boldsymbol{j}-10\boldsymbol{k}.\end{aligned}$$

■

Parallel (collinear) vectors

Two vectors $\boldsymbol{a}$ and $\boldsymbol{b}$ are parallel or collinear if there is a constant c such that $\boldsymbol{a}=c\boldsymbol{b}$. If $\boldsymbol{a}=a_1\boldsymbol{i}+a_2\boldsymbol{j}+a_3\boldsymbol{k}$ and $\boldsymbol{b}=b_1\boldsymbol{i}+b_2\boldsymbol{j}+b_3\boldsymbol{k}$, this leads to

$$a_1=cb_1,\ a_2=cb_2,\ a_3=cb_3\Rightarrow\frac{a_1}{b_1}=\frac{a_2}{b_2}=\frac{a_3}{b_3}.$$

That is, two vectors are parallel or collinear if and only if their components are proportional. Note that a denominator can be 0, with the convention that the corresponding numerator is also 0.

■ **Example 6.6** Find x, y such that the vectors $\boldsymbol{a}=\boldsymbol{i}+(x-y)\boldsymbol{j}+2\boldsymbol{k}$ and $\boldsymbol{b}=-\boldsymbol{i}+2\boldsymbol{j}+y\boldsymbol{k}$ are collinear. Since $\boldsymbol{a}$ and $\boldsymbol{b}$ are collinear, their components are proportional. It follows that

$$\frac{1}{-1}=\frac{x-y}{2}=\frac{2}{y}\Rightarrow-1=\frac{x-y}{2}=\frac{2}{y}\Rightarrow x-y=-2,\ y=-2\Rightarrow x=-4,y=-2.$$

■

The position vector

To every point $P(a,b,c)$ one can uniquely associate its **position vector** $\overrightarrow{OP}$, which characterizes the position of P with respect to the origin O. The position vector can also be denoted $\boldsymbol{r}_P$ and is defined by

$$\overrightarrow{OP}=a\boldsymbol{i}+b\boldsymbol{j}+c\boldsymbol{k},$$

multiplying the components of P with the vectors of the standard basis and adding up the results.

■ **Example 6.7** The position vector of $A(3,2,-5)$ is $\overrightarrow{OA}=3\boldsymbol{i}+2\boldsymbol{j}-5\boldsymbol{k}$.

■

■ **Example 6.8** If the position vector of B is $\overrightarrow{OB}=2\boldsymbol{i}+5\boldsymbol{j}+6\boldsymbol{k}$, then $B=B(2,5,6)$.

■

The vector between two points

Let $A(x_1, y_1, z_1)$ and $B(x_2, y_2, z_2)$ be two points. Then the vector $\overrightarrow{AB}$ has the expression

$$\begin{aligned}\overrightarrow{AB}&=\overrightarrow{OB}-\overrightarrow{OA}\\&=(x_2\boldsymbol{i}+y_2\boldsymbol{j}+z_2\boldsymbol{k})-(x_1\boldsymbol{i}+y_1\boldsymbol{j}+z_1\boldsymbol{k})\\&=(x_2-x_1)\boldsymbol{i}+(y_2-y_1)\boldsymbol{j}+(z_2-z_1)\boldsymbol{k}.\end{aligned}$$

® Note again the mnemonics "arrival minus departure". That is, the coordinates of $\overrightarrow{AB}$, vector which can also be represented as

$$\overrightarrow{AB}=(x_2-x_1, y_2-y_1, z_2-z_1),$$

are obtained by subtracting the coordinates of the "departure" (initial) point, A, from the coordinates of the "arrival" (terminal) point, B.

■ **Example 6.9** Given the points $A(4,-1,3)$ and $B(-2,6,5)$, find the vector $\overrightarrow{AB}$.

We subtract the coordinates of the initial point, A, from the coordinates of the terminal point, B, and obtain

$$\overrightarrow{AB}=(-2-4, 6-(-1), 5-3)=(-6,7,2).$$

Equivalently, $\overrightarrow{AB}=-6\boldsymbol{i}+7\boldsymbol{j}+2\boldsymbol{k}$.

■

6.2.2 Vectors in two dimensions

It is entirely possible to state similar algebraic considerations for vectors in a plane, rather than for vectors in space. In this case, the standard basis will be $\{\boldsymbol{i},\boldsymbol{j}\}$ (no $\boldsymbol{k}$, that is). We shall briefiy review the main points.

Any vector $\boldsymbol{a}$ has two components a_1, a_2, which are uniquely determined, such that

$$\boldsymbol{a}=a_1\boldsymbol{i}+a_2\boldsymbol{j},$$

and, in this context, the vector $\boldsymbol{a}$ may also represented by indicating its components, in the form

$$\boldsymbol{a}=(a_1, a_2).$$

If $\boldsymbol{a}=a_1\boldsymbol{i}+a_2\boldsymbol{j}$, then the magnitude (length) of $\boldsymbol{a}$ is

$$\|\boldsymbol{a}\|=\sqrt{a_1^2+a_2^2}.$$

Exercise 6.2 Find the length of the vector $\boldsymbol{v}=-5\boldsymbol{i}+10\boldsymbol{j}$.

■

Solution Using the length formula, we obtain

$$\begin{aligned}\|\boldsymbol{v}\|&=\sqrt{(-5)^2+(10)^2}\\&=\sqrt{25+100}\\&=\sqrt{125}\\&=5\sqrt{5}.\end{aligned}$$

■

Two vectors are equal if and only if their components are equal.

Exercise 6.3 If $\boldsymbol{p}=(a+3, b-1)$ and $\boldsymbol{q}=(-2, 3)$, find a and b such that $\boldsymbol{p}=\boldsymbol{q}$. ■

Solution Since the vectors $\boldsymbol{p}$ and $\boldsymbol{q}$ are equal, they need to have equal components. It follows that

$$a+3=-2 \Rightarrow a=-5.$$
$$b-1=3 \Rightarrow b=4.$$

■

As previously mentioned, addition and subtraction of vectors reduce to addition and subtraction, respectively, of their components. Scalar multiplication of a vector reduce to the multiplication of its components with the scalar.

Exercise 6.4 If $\boldsymbol{u}=2\boldsymbol{i}+3\boldsymbol{j}$ and $\boldsymbol{v}=3\boldsymbol{i}-2\boldsymbol{j}$, find $3\boldsymbol{u}+2\boldsymbol{v}$. ■

Solution

$$\begin{aligned} 3\boldsymbol{u}+2\boldsymbol{v} &= 3(2\boldsymbol{i}+3\boldsymbol{j})+2(3\boldsymbol{i}-2\boldsymbol{j}) \\ &= (6\boldsymbol{i}+9\boldsymbol{j})+(6\boldsymbol{i}-4\boldsymbol{j}) \\ &= 12\boldsymbol{i}+5\boldsymbol{j}. \end{aligned}$$

■

Two vectors are parallel or collinear if and only if their components are proportional.

Exercise 6.5 Find a such that the vectors $\boldsymbol{u}=(2a-3)\boldsymbol{i}-(a-4)\boldsymbol{j}$ and $\boldsymbol{v}=(a+1)\boldsymbol{i}+6\boldsymbol{j}$ are collinear. ■

Solution Since the vectors are collinear, their components are proportional. It follows that

$$\frac{2a-3}{a+1}=\frac{-(a-4)}{6},$$
$$6(2a-3)=-(a+1)(a-4),$$
$$12a-18=-a^2+3a+4,$$
$$a^2+9a-22=0.$$

Solving the second degree equation, one obtains that $a=2$ or $a=-11$.

■

Exercise 6.6 Let $\boldsymbol{a}=3\boldsymbol{i}+4\boldsymbol{j}$. Find a vector of length 6 collinear with $\boldsymbol{a}$. ■

Solution We search for $\boldsymbol{b}=c\boldsymbol{a}$, where c is a constant. Then $\boldsymbol{b}=3c\boldsymbol{i}+4c\boldsymbol{j}$ and consequently

$$\begin{aligned} \|\boldsymbol{b}\| &= \sqrt{(3c)^2+(4c)^2} \\ &= \sqrt{9c^2+16c^2} \end{aligned}$$

$$= \sqrt{25c^2}$$
$$= 5|c|.$$

Then $5|c| = 6$ and consequently $c = \pm\frac{6}{5}$. The first solution is $\boldsymbol{b}_1 = \frac{6}{5}(3\boldsymbol{i} + 4\boldsymbol{j})$ and the second solution is $\boldsymbol{b}_2 = -\frac{6}{5}(3\boldsymbol{i} + 4\boldsymbol{j})$.

■

6.3 The scalar product

The scalar or dot product of two non-zero vectors $\boldsymbol{a}$ and $\boldsymbol{b}$ is defined as

$$\boldsymbol{a} \cdot \boldsymbol{b} = \|\boldsymbol{a}\| \|\boldsymbol{b}\| \cos\theta,$$

where θ is the angle between $\boldsymbol{a}$ and $\boldsymbol{b}$. If any of $\boldsymbol{a}$ and $\boldsymbol{b}$ are $\boldsymbol{0}$, the scalar product is defined as being 0.

Ⓡ The dot (scalar) product of two vectors is a **number** (scalar). That is why it is called a scalar product.

Ⓡ Suppose that $\boldsymbol{a}$ and $\boldsymbol{b}$ are non-zero. Then θ can be an acute, right or obtuse angle.

- If θ is an acute angle, then $\boldsymbol{a} \cdot \boldsymbol{b}$ is positive (because the cosine of an acute angle is positive).
- If θ is a right angle, then $\boldsymbol{a} \cdot \boldsymbol{b}$ is 0 (because the cosine of a right angle is 0).
- If θ is an obtuse angle, then $\boldsymbol{a} \cdot \boldsymbol{b}$ is negative (because the cosine of an obtuse angle is negative).

Ⓡ If $\boldsymbol{a}$ and $\boldsymbol{b}$ are codirectional, i.e. $\theta = 0$, then $\boldsymbol{a} \cdot \boldsymbol{b} = \|\boldsymbol{a}\| \|\boldsymbol{b}\|$ because of $\cos 0 = 1$. This implies that the dot product of a vector $\boldsymbol{a}$ with itself is $\boldsymbol{a} \cdot \boldsymbol{a} = \|\boldsymbol{a}\|^2$, which gives $\|\boldsymbol{a}\| = \sqrt{\boldsymbol{a} \cdot \boldsymbol{a}}$.

Ⓡ The scalar products of vectors in the standard basis are as follows

$$\boldsymbol{i} \cdot \boldsymbol{i} = \|\boldsymbol{i}\|^2 = 1;$$

$$\boldsymbol{i} \cdot \boldsymbol{j} = \|\boldsymbol{i}\| \|\boldsymbol{j}\| \cos\frac{\pi}{2} = 0;$$

$$\boldsymbol{j} \cdot \boldsymbol{i} = \|\boldsymbol{j}\| \|\boldsymbol{i}\| \cos\frac{\pi}{2} = 0;$$

$$\boldsymbol{j} \cdot \boldsymbol{j} = \|\boldsymbol{j}\|^2 = 1.$$

6.3.1 The algebraic expression

If $\boldsymbol{a} = (x_1, x_2, x_3)$ and $\boldsymbol{b} = (y_1, y_2, y_3)$, then in terms of components the dot product of $\boldsymbol{a}$ and $\boldsymbol{b}$ is

$$\boldsymbol{a} \cdot \boldsymbol{b} = x_1 y_1 + x_2 y_2 + x_3 y_3.$$

Similarly, if $\boldsymbol{a} = (x_1, x_2)$ and $\boldsymbol{b} = (y_1, y_2)$, then

$$\boldsymbol{a} \cdot \boldsymbol{b} = x_1 y_1 + x_2 y_2.$$

Exercise 6.7 If $\boldsymbol{a} = (2, 1)$ and $\boldsymbol{b} = (3, -2)$, find $\boldsymbol{a} \cdot \boldsymbol{b}$.

■

Solution $\boldsymbol{a} \cdot \boldsymbol{b}=(2\times 3)+(1\times(-2))=4$.

■

Exercise 6.8 If $\boldsymbol{m}=(-2,3,1)$ and $\boldsymbol{n}=(1,-5,2)$, find $\boldsymbol{m} \cdot \boldsymbol{n}$.

■

Solution $\boldsymbol{m} \cdot \boldsymbol{n}=(-2\times 1)+(3\times(-5))+(1\times 2)=-15$.

■

Exercise 6.9 If $\boldsymbol{a}=4\boldsymbol{i}+3\boldsymbol{j}$ and $\boldsymbol{b}=-5\boldsymbol{i}+2\boldsymbol{j}$, find $\boldsymbol{a} \cdot \boldsymbol{b}$.

■

Solution $\boldsymbol{a} \cdot \boldsymbol{b}=(4\boldsymbol{i}+3\boldsymbol{j}) \cdot (-5\boldsymbol{i}+2\boldsymbol{j})=-20+6=-14$.

■

6.3.2 Properties of the dot product

(1) Commutative: $\boldsymbol{a} \cdot \boldsymbol{b}=\boldsymbol{b} \cdot \boldsymbol{a}$.

(2) Distributive over vector addition: $\boldsymbol{a} \cdot (\boldsymbol{b}+\boldsymbol{c})=\boldsymbol{a} \cdot \boldsymbol{b}+\boldsymbol{a} \cdot \boldsymbol{c}$.

(3) Bilinear: $\boldsymbol{a} \cdot (r\boldsymbol{b}+\boldsymbol{c})=r(\boldsymbol{a} \cdot \boldsymbol{b})+(\boldsymbol{a} \cdot \boldsymbol{c})$.

(4) Scalar multiplication: $(c_1\boldsymbol{a}) \cdot (c_2\boldsymbol{b})=c_1c_2(\boldsymbol{a} \cdot \boldsymbol{b})$.

(5) Not associative because the dot product between a scalar $(\boldsymbol{a} \cdot \boldsymbol{b})$ and a vector $(\boldsymbol{c})$ is not defined.

(6) Orthogonal: Two non-zero vectors $\boldsymbol{a}$ and $\boldsymbol{b}$ are orthogonal if and only if $\boldsymbol{a} \cdot \boldsymbol{b}=0$.

(7) No cancellation: Unlike multiplication of ordinary numbers, where if $ab=ac$, then b always equals c unless a is zero, the dot product does not obey the cancellation law.

6.3.3 Applications of the dot product

The angle between two vectors

From the definition of dot product, we obtain

$$\cos\theta=\frac{\boldsymbol{a} \cdot \boldsymbol{b}}{\|\boldsymbol{a}\|\|\boldsymbol{b}\|}.$$

where θ is the angle between the vectors $\boldsymbol{a}$ and $\boldsymbol{b}$.

$$\cos\theta=\frac{\boldsymbol{a}}{\|\boldsymbol{a}\|} \cdot \frac{\boldsymbol{b}}{\|\boldsymbol{b}\|}=\boldsymbol{a} \cdot \boldsymbol{b},$$

Therefore the angle between the vectors $\boldsymbol{a}$ and $\boldsymbol{b}$ is the angle between their unit vectors $\boldsymbol{a}$ and $\boldsymbol{b}$.

Exercise 6.10 Find the angle between the vectors $\boldsymbol{p}=(-\sqrt{3},1)$ and $\boldsymbol{q}=(0,-3)$.

■

Solution Denote by θ be the angle between the vectors. By the definition of dot product,

$$\boldsymbol{p} \cdot \boldsymbol{q}=-\sqrt{3}\times 0+1\times(-3)=-3,$$

$$\|\boldsymbol{p}\|=\sqrt{(-\sqrt{3})^2+1^2}=2,$$

$$\| \boldsymbol{q} \| = \sqrt{0^2 + (-3)^2} = 3,$$

$$\cos\theta = \frac{\boldsymbol{p} \cdot \boldsymbol{q}}{\| \boldsymbol{p} \| \cdot \| \boldsymbol{q} \|} = \frac{-3}{2\times 3} \Rightarrow \theta = \frac{2\pi}{3}.$$

■

The law of cosines

Given a triangle with vector edges $\boldsymbol{a}$ and $\boldsymbol{b}$, separated by the angle θ (see Figure 6.9), the third side is $\boldsymbol{c}=\boldsymbol{a}-\boldsymbol{b}$.

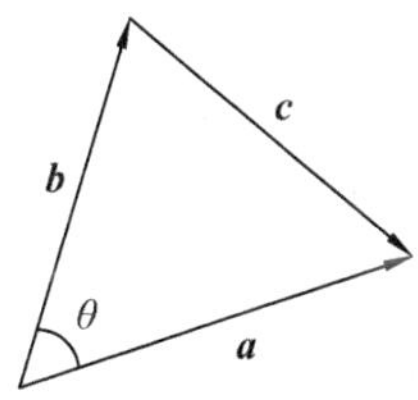

Figure 6.9 Triangle with vector edges a and b, separated by the angle θ.

The dot product of $\boldsymbol{c}$ with itself is

$$\begin{aligned} \boldsymbol{c} \cdot \boldsymbol{c} &= (\boldsymbol{a}-\boldsymbol{b}) \cdot (\boldsymbol{a}-\boldsymbol{b}) \\ &= \boldsymbol{a} \cdot \boldsymbol{a} - \boldsymbol{a} \cdot \boldsymbol{b} - \boldsymbol{b} \cdot \boldsymbol{a} + \boldsymbol{b} \cdot \boldsymbol{b} \\ &= \| \boldsymbol{a} \|^2 - 2\boldsymbol{a} \cdot \boldsymbol{b} + \| \boldsymbol{b} \|^2 \\ &= \| \boldsymbol{a} \|^2 + \| \boldsymbol{b} \|^2 - 2 \| \boldsymbol{a} \| \| \boldsymbol{b} \| \cos\theta. \end{aligned}$$

Hence

$$\| \boldsymbol{c} \|^2 = \| \boldsymbol{a} \|^2 + \| \boldsymbol{b} \|^2 - 2 \| \boldsymbol{a} \| \| \boldsymbol{b} \| \cos\theta,$$

which is the law of cosines.

6.4 The vector product

The vector or cross product of two vectors $\boldsymbol{a}$ and $\boldsymbol{b}$ is defined only in the three-dimensional space and is given by

$$\boldsymbol{a} \times \boldsymbol{b} = \| \boldsymbol{a} \| \| \boldsymbol{b} \| \sin\theta\, \boldsymbol{n},$$

where θ is the angle between $\boldsymbol{a}$ and $\boldsymbol{b}$ in the plane containing them (it is between 0° and 180°) and $\boldsymbol{n}$ is a unit vector perpendicular to the plane containing $\boldsymbol{a}$ and $\boldsymbol{b}$ in the direction given by the right-hand rule (the vector $\boldsymbol{n}$ is coming out of the thumb) as shown in Figure 6.10.

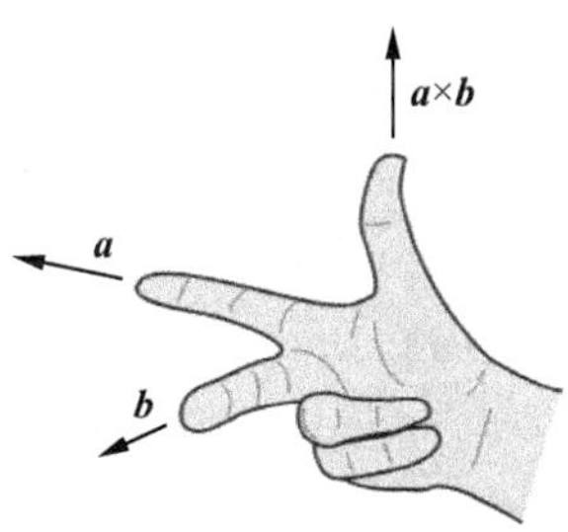

Figure 6.10 The direction of the cross product by the right-hand rule.

Ⓡ The cross (vector) product of two vectors is a **vector**. That is why it is called a vector product.

Ⓡ The use of the right-hand rule implies that the cross-product is anti-commutative, i.e.,

$$\boldsymbol{b}\times\boldsymbol{a}=-(\boldsymbol{a}\times\boldsymbol{b}).$$

By pointing the forefinger toward $\boldsymbol{b}$ first, and then pointing the middle finger toward $\boldsymbol{a}$, the thumb will be forced in the opposite direction, reversing the direction of the product vector.

Ⓡ If a left-handed coordinate system is used, the direction of the vector $\boldsymbol{n}$ is given by the left-hand rule and points in the opposite direction.

Products of vectors in the standard basis

The standard basis vectors $\boldsymbol{i}$, $\boldsymbol{j}$ and $\boldsymbol{k}$ (see Figure 6.8) satisfy the following equalities in a right hand coordinate system:

$$\boldsymbol{i}\times\boldsymbol{j}=\boldsymbol{k},$$
$$\boldsymbol{j}\times\boldsymbol{k}=\boldsymbol{i},$$
$$\boldsymbol{k}\times\boldsymbol{i}=\boldsymbol{j},$$

which imply, by the anti-commutativity of the cross product, that

$$\boldsymbol{j}\times\boldsymbol{i}=-\boldsymbol{k},$$
$$\boldsymbol{k}\times\boldsymbol{j}=-\boldsymbol{i},$$
$$\boldsymbol{i}\times\boldsymbol{k}=-\boldsymbol{j}.$$

The definition of the cross product also implies that

$$\boldsymbol{i}\times\boldsymbol{i}=\boldsymbol{j}\times\boldsymbol{j}=\boldsymbol{k}\times\boldsymbol{k}=\boldsymbol{0}.$$

Ⓡ Notice the mnemonics: if the vectors for which the cross product is computed appear in the right order in the sequence

$$\boldsymbol{i}\to\boldsymbol{j}\to\boldsymbol{k}\to\boldsymbol{i}\to\boldsymbol{j},$$

then the product is the next vector in the sequence. If not, then it's its opposite. For instance, $\boldsymbol{i}\times\boldsymbol{j}=\boldsymbol{k}$, since $\boldsymbol{i}$, $\boldsymbol{j}$ appear in that order in the sequence and the next vector in the sequence is $\boldsymbol{k}$. However, $\boldsymbol{j}\times\boldsymbol{i}=-\boldsymbol{k}$, since $\boldsymbol{j}$, $\boldsymbol{i}$ do not appear in that order in the sequence, but in the order $\boldsymbol{i}$, $\boldsymbol{j}$, the next vector being $\boldsymbol{k}$.

6.4.1 The algebraic expression

The distributivity and linearity of the cross product are sufficient to determine the cross product of any two vectors $\boldsymbol{u}$ and $\boldsymbol{v}$. Each vector can be defined as the sum of three orthogonal components parallel to the standard basis vectors

$$\boldsymbol{u}=u_1\boldsymbol{i}+u_2\boldsymbol{j}+u_3\boldsymbol{k} \quad\text{and}\quad \boldsymbol{v}=v_1\boldsymbol{i}+v_2\boldsymbol{j}+v_3\boldsymbol{k}.$$

Their cross product $\boldsymbol{u}\times\boldsymbol{v}$ can be expanded using distributivity

$$\begin{aligned}\boldsymbol{u}\times\boldsymbol{v}&=(u_1\boldsymbol{i}+u_2\boldsymbol{j}+u_3\boldsymbol{k})\times(v_1\boldsymbol{i}+v_2\boldsymbol{j}+v_3\boldsymbol{k})\\&=u_1v_1(\boldsymbol{i}\times\boldsymbol{i})+u_1v_2(\boldsymbol{i}\times\boldsymbol{j})+u_1v_3\boldsymbol{i}\times\boldsymbol{k}+u_2v_1(\boldsymbol{j}\times\boldsymbol{i})+u_2v_2(\boldsymbol{j}\times\boldsymbol{j})+\end{aligned}$$

$$u_2 v_3 \boldsymbol{j} \times \boldsymbol{k} + u_3 v_1 (\boldsymbol{k} \times \boldsymbol{i}) + u_3 v_2 (\boldsymbol{k} \times \boldsymbol{j}) + u_3 v_3 \boldsymbol{k} \times \boldsymbol{k}$$
$$= (u_2 v_3 - u_3 v_2)\boldsymbol{i} + (u_3 v_1 - u_1 v_3)\boldsymbol{j} + (u_1 v_2 - u_2 v_1)\boldsymbol{k}.$$

Exercise 6.11 Let $\boldsymbol{a} = 2\boldsymbol{i} - 2\boldsymbol{j} + 1\boldsymbol{k}$ and $\boldsymbol{b} = -5\boldsymbol{i} + 4\boldsymbol{j} + 7\boldsymbol{k}$. Compute $\boldsymbol{a} \times \boldsymbol{b}$. ■

Solution

$$\boldsymbol{a} \times \boldsymbol{b} = ((-2) \cdot 7 - 1 \cdot 4)\boldsymbol{i} + (1 \cdot (-5) - 2 \cdot 7)\boldsymbol{j} + (2 \cdot 4 - (-2) \cdot (-5))\boldsymbol{k}$$
$$= -18\boldsymbol{i} - 19\boldsymbol{j} - 2\boldsymbol{k}.$$

■

6.4.2 Geometric applications

The area of a parallelogram

The magnitude of the cross product can be interpreted as the positive area of the parallelogram having $\boldsymbol{a}$ and $\boldsymbol{b}$ as sides, as seen in Figure 6.11.

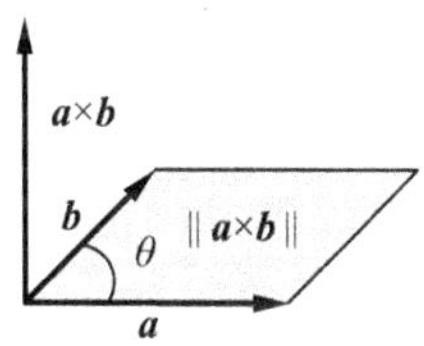

Figure 6.11 The area of a parallelogram as the magnitude of a cross product.

$$\| \boldsymbol{a} \times \boldsymbol{b} \| = \| \boldsymbol{a} \| \| \boldsymbol{b} \| \sin \theta.$$

Ⓡ If $\boldsymbol{a} \neq \boldsymbol{0}$ and $\boldsymbol{b} \neq \boldsymbol{0}$, then

$$\theta = \sin^{-1}\left(\frac{\| \boldsymbol{a} \times \boldsymbol{b} \|}{\| \boldsymbol{a} \| \| \boldsymbol{b} \|} \right),$$

in which $\theta \in [0, \pi]$.

The volume of a parallelepiped

One can also compute the volume V of a parallelepiped having $\boldsymbol{a}$, $\boldsymbol{b}$ and $\boldsymbol{c}$ as edges by using a combination of a cross product and a dot product, called scalar triple product (see Figure 6.12)

$$\boldsymbol{a} \cdot (\boldsymbol{b} \times \boldsymbol{c}) = \boldsymbol{b} \cdot (\boldsymbol{c} \times \boldsymbol{a}) = \boldsymbol{c} \cdot (\boldsymbol{a} \times \boldsymbol{b}).$$

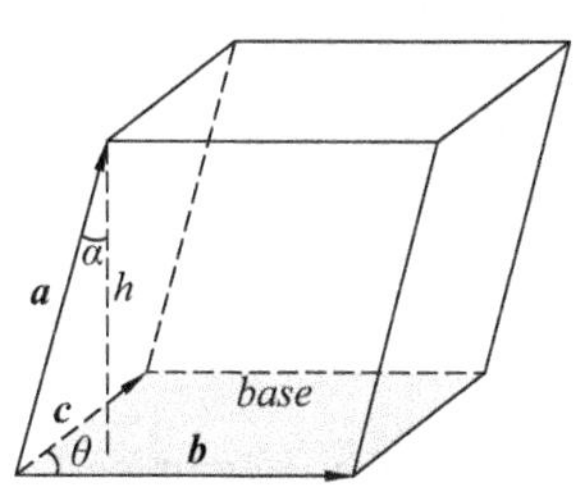

Figure 6.12 Three vectors defining a parallelepiped.

Since the result of the scalar triple product may be negative, the volume of the parallelepiped is given by its absolute value

$$V = \| \boldsymbol{a} \cdot (\boldsymbol{b} \times \boldsymbol{c}) \|.$$

Exercise 6.12 Find the area of the parallelogram spanned by $\boldsymbol{a}=2\boldsymbol{i}-2\boldsymbol{j}+1\boldsymbol{k}$ and $\boldsymbol{b}=-5\boldsymbol{i}+4\boldsymbol{j}+7\boldsymbol{k}$. ■

Solution In the last exercise we found that

$$\boldsymbol{a} \times \boldsymbol{b} = -18\boldsymbol{i} - 19\boldsymbol{j} - 2\boldsymbol{k}.$$

Therefore, the area of the parallelogram is

$$\| \boldsymbol{a} \times \boldsymbol{b} \| = \sqrt{(-18)^2+(-19)^2+(-2)^2} = \sqrt{689}.$$

■

Exercise 6.13 Compute the volume V of a parallelepiped having $\boldsymbol{a}=2\boldsymbol{i}-2\boldsymbol{j}+1\boldsymbol{k}$, $\boldsymbol{b}=-5\boldsymbol{i}+4\boldsymbol{j}+7\boldsymbol{k}$ and $\boldsymbol{c}=\boldsymbol{i}-\boldsymbol{j}+2\boldsymbol{k}$ as edges ■

Solution Since $\boldsymbol{a} \times \boldsymbol{b} = -18\boldsymbol{i} - 19\boldsymbol{j} - 2\boldsymbol{k}$, the volume is

$$V = \| \boldsymbol{c} \cdot (\boldsymbol{a} \times \boldsymbol{b}) \| = |1 \cdot (-18) + (-1) \cdot (-19) + 2 \cdot (-2)| = 3.$$

■

Direction cosines

Suppose we have a vector $\overrightarrow{OA}$ with initial point at the origin and terminal point at A. Suppose also that we have a unit vector in the same direction as $\overrightarrow{OA}$ and let the unit vector be

$$\boldsymbol{u} = \mu_1 \boldsymbol{i} + \mu_2 \boldsymbol{j} + \mu_3 \boldsymbol{k}.$$

As shown in the left graph of Figure 6.13, $\boldsymbol{u}$ is the unit vector pointing in the same direction as the vector $\overrightarrow{OA}$, and $\boldsymbol{i}$, $\boldsymbol{j}$ and $\boldsymbol{k}$ (the unit vectors in the x-, y- and z-directions respectively) are marked in boldface type.

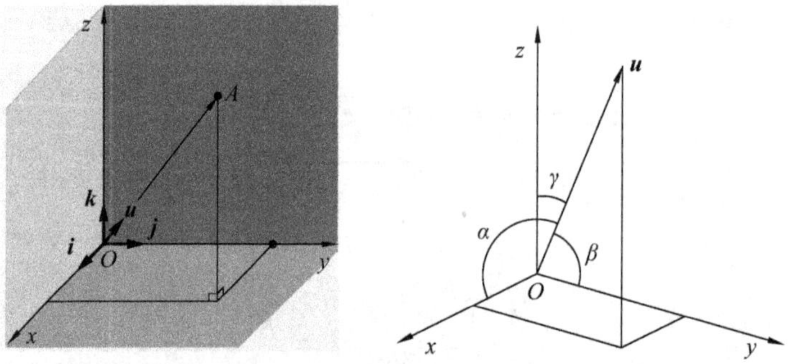

Figure 6.13 A vector in a 3-dimensional Euclidean space.

Focus now on the vector $\boldsymbol{u}$. If in the right diagram of Figure 6.13, α is the angle between $\boldsymbol{u}$ and the semi-axis Ox (in dark red), β is the angle between $\boldsymbol{u}$ and the semi-axis Oy (in green) and γ is the angle between u and the semi-axis Oz (in pink), then we can use the scalar product and obtain that

$$\mu_1 = \boldsymbol{u} \cdot \boldsymbol{i} = 1 \times 1 \times \cos \alpha = \cos \alpha;$$

$$\mu_2 = \boldsymbol{u} \cdot \boldsymbol{j} = 1 \times 1 \times \cos \beta = \cos \beta;$$

$$\mu_3 = \boldsymbol{u} \cdot \boldsymbol{k} = 1 \times 1 \times \cos \gamma = \cos \gamma.$$

Consequently, the unit vector $\boldsymbol{u}$ is given by

$$\boldsymbol{u} = \cos \alpha \boldsymbol{i} + \cos \beta \boldsymbol{j} + \cos \gamma \boldsymbol{k}.$$

These 3 cosines are called the **direction cosines.**

6.5 Equations of a straight line

6.5.1 The vector equation of a straight line

We are already familiar with the equation of a line in two dimensions: the line with slope m and y-intercept c has equation

$$y = mx + c.$$

The equation can be deduced without vectors, but vectors provide a really clear and quick way into the problem. We need to know two things: a point through which the line passes, and the direction of the line. Both of them can be described using vectors.

Figure 6.14 pictures the straight line through the point A (with position vector $\boldsymbol{a}$), parallel to the vector $\boldsymbol{d}$. How do we get from the origin to a general point P on the line where P has position vector $\boldsymbol{r}$? One answer is that we first get to the point A, by travelling along the vector $\boldsymbol{a}$, and then travel a certain distance in the direction of the vector $\boldsymbol{d}$. If the position vector of P is $\boldsymbol{r}$, this implies that there is a value of λ such that

$$\boldsymbol{r} = \boldsymbol{a} + \lambda \boldsymbol{d}.$$

The above equation is called **the vector equation of a straight line.**

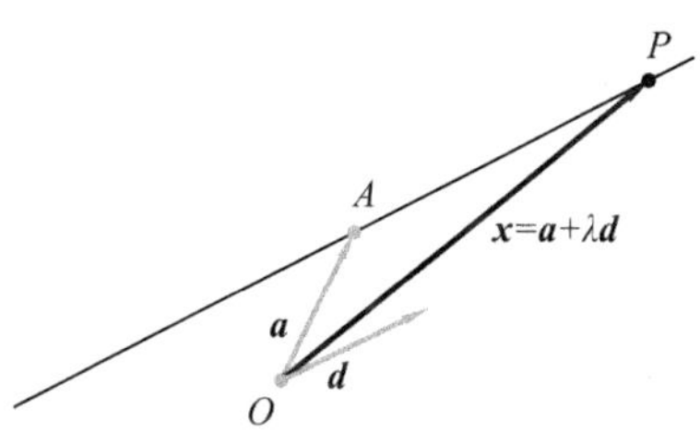

Figure 6.14 A straight line through the point A (with position vector $\boldsymbol{a}$), parallel to the vector $\boldsymbol{d}$.

Ⓡ The vector equation of a straight line passing through the origin and parallel to a given vector $\boldsymbol{d}$ will be of the form $\boldsymbol{r} = \lambda \boldsymbol{d}$.

■ **Example 6.10** The line through the point $(2, 0, -1)$ parallel to the vector $-3\boldsymbol{i} + \boldsymbol{j} - \boldsymbol{k}$ has equation

$$\boldsymbol{r} = 2\boldsymbol{i} - \boldsymbol{k} + \lambda(-3\boldsymbol{i} + \boldsymbol{j} - \boldsymbol{k}).$$

■

6.5.2 The standard Cartesian equation of a straight line

Identifying components for the vectors in each side of the vector equation of the straight line passing through $P_1(x_1, y_1, z_1)$ and parallel to $\boldsymbol{d} = d_1\boldsymbol{i} + d_2\boldsymbol{j} + d_3\boldsymbol{j}$ leads to the parametric equations

$$x=x_1+d_1\lambda\ ,\ y=y_1+d_2\lambda,\ z=z_1+d_3\lambda,$$

and if these are solved for the parameter $\lambda\in\mathbf{R}$, we obtain

$$\frac{x-x_1}{d_1}=\frac{y-y_1}{d_2}=\frac{z-z_1}{d_3}.$$

This equation is called **the standard Cartesian equation of a straight line.**

Ⓡ Notice the mnemonics: the coordinates of the point go above (at the numerator), with a change of sign, the components of the vector go below (at the denominator).

Exercise 6.14 Determine the vector equation of the straight line passing through the point with position vector $\boldsymbol{a}=\boldsymbol{i}-3\boldsymbol{j}+\boldsymbol{k}$ which is parallel to the vector $\boldsymbol{d}=2\boldsymbol{i}+3\boldsymbol{j}-4\boldsymbol{k}$. Express the vector equation of the straight line in standard Cartesian form. ■

Solution The vector equation of the straight line is

$$\boldsymbol{r}=\boldsymbol{i}-3\boldsymbol{j}+\boldsymbol{k}+\lambda(2\boldsymbol{i}+3\boldsymbol{j}-4\boldsymbol{k})$$

or

$$x\boldsymbol{i}+y\boldsymbol{j}+z\boldsymbol{k}=(1+2\lambda)\boldsymbol{i}+(-3+3\lambda)\boldsymbol{j}+(1-4\lambda)\boldsymbol{k}.$$

Eliminating λ from each component, we obtain the Cartesian form of the straight line

$$\frac{x-1}{2}=\frac{y+3}{3}=\frac{z-1}{-4}.$$

■

Exercise 6.15 The equation

$$\frac{3x+1}{2}=\frac{y-1}{2}=\frac{-z+5}{3}$$

determines a straight line. Express this equation in vector form. ■

Solution Rewriting the equation so that the coefficients of x, y and z are 1, we have

$$\frac{x+\frac{1}{3}}{\frac{2}{3}}=\frac{y-1}{2}=\frac{z-5}{-3}.$$

Hence, in vector form, the equation of the line is

$$\boldsymbol{r}=-\frac{1}{3}\boldsymbol{i}+\boldsymbol{j}+5\boldsymbol{k}+\lambda\left(\frac{2}{3}\boldsymbol{i}+2\boldsymbol{j}-3\boldsymbol{k}\right).$$

■

Exercise 6.16 Show that the two straight lines $\boldsymbol{r}=\boldsymbol{r}_1+t\boldsymbol{a}_1$ and $\boldsymbol{r}=\boldsymbol{r}_2+t\boldsymbol{a}_2$, where $\boldsymbol{r}_1=\boldsymbol{j}$, $\boldsymbol{a}_1=\boldsymbol{i}+2\boldsymbol{j}-\boldsymbol{k}$ and $\boldsymbol{r}_2=\boldsymbol{i}+\boldsymbol{j}+\boldsymbol{k}$, $\boldsymbol{a}_2=-2\boldsymbol{i}-2\boldsymbol{j}$, have a common point and determine its co-ordinates. ■

Solution Any point on the first line is such that $x=t$, $y=1+2t$, $z=-t$, for some parameter value, $t\in\mathbf{R}$. Also, any point on the second line is such that $x=1-2s$, $y=1-2s$, $z=1$, for some parameter value, $s\in\mathbf{R}$ (notice that, since we write equations for two distinct lines, we need to use two distinct parameters s and t; using only one would be wrong).

The lines have a common point if it is possible to find values of t and s such that $t=1-2s, 1+2t=1-2s$ and $-t=1$. We obtain that $t=-1$ and $s=1$, which give the common point $P(-1,-1,1)$.

■

6.6 Equations of a plane

6.6.1 The vector equation of a plane

Let us start by assuming that we know a point that is on the plane, $P_1(x_1, y_1, z_1)$. Let us also suppose that there is a vector that is orthogonal (perpendicular) to the plane, $\boldsymbol{n}=a\boldsymbol{i}+b\boldsymbol{j}+c\boldsymbol{k}$. This vector is called the **normal vector.** Now, assume that $P(x,y,z)$ is an arbitrary point on the plane. Finally, since we shall be going to be work primarily with vectors, we will let $\boldsymbol{p}_1$ and $\boldsymbol{p}$ be the position vectors for points P_1 and P, respectively.

Figure 6.15 shows that the vector $\boldsymbol{p}-\boldsymbol{p}_1$ and the normal vector $\boldsymbol{n}$ are perpendicular to each other, which implies that their dot product should be 0. That is,

$$(\boldsymbol{p}-\boldsymbol{p}_1)\cdot\boldsymbol{n}=0.$$

The above equation is called **the vector equation of a plane.**

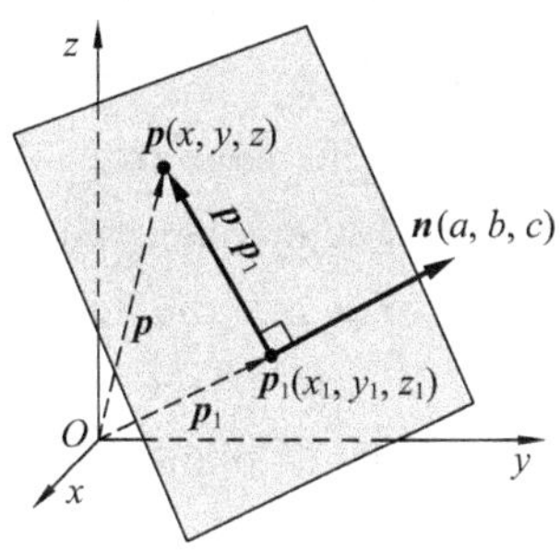

Figure 6.15 Graph of a plane in 3D.

6.6.2 The standard Cartesian equation of a plane

Identifying components for each vector appearing in the above vector equation implies that

$$(x-x_1, y-y_1, z-z_1)(a,b,c)=0.$$

This leads to

$$a(x-x_1)+b(y-y_1)+c(z-z_1)=0.$$

If we denote $d=(ax_1+by_1+cz_1)$, then the equation of the plane becomes

$$ax+by+cz=d.$$

Ⓡ Note that the coefficients a, b, c of x, y and z, respectively, are the components of the normal vector $\boldsymbol{n}$.

Exercise 6.17 Determine the equation of the plane that contains the points

$$P=(1,1,1),\ Q=(1,2,0),\ R=(-1,2,1).$$

Solution We seek for the coefficients of an equation $ax+by+cz=d$ satisfied by the components of P, Q and R. Thus we have

$$a+b+c=d,$$
$$a+2b+0c=d,$$
$$-a+2b+c=d.$$

Subtracting the first equation from the second and then adding the first equation to the third, we eliminate a to get

$$b-c=0,$$
$$4b+c=2d.$$

Adding the equations gives $5b=2d$ or $b=\frac{2d}{5}$, then solving for c leads to $c=b=\frac{2d}{5}$, which implies that $a=d-b-c=\frac{d}{5}$. Consequently, the equation (obtained for $d=1$) is

$$x+2y+2z=5 \text{ or } \frac{x}{5}+\frac{2y}{5}+\frac{2z}{5}=1.$$

6.6.3 The distance from the origin to a given plane

If the normal vector is taken to be of length 1 (normalized), then the corresponding constant term of the plane equation is the distance from the origin to the plane.

To prove this, let us denote by D the distance from the origin to the plane. As shown in Figure 6.16, if the unit normal vector $\boldsymbol{n}$ has the coordinates $\boldsymbol{n}_1=(a_1,b_1,c_1)$, then the point P_1 in which the perpendicular from 0 meets the plane has coordinates $P_1=(Da_1,Db_1,Dc_1)$. If $P(x,y,z)$ is an arbitrary point in the plane, then the equation of the plane can be written as

$$(a_1,b_1,c_1)\cdot(x-Da_1,y-Db_1,z-Dc_1)=0,$$

which indicates that

$$a_1x+b_1y+c_1z-D(a_1^2+b_1^2+c_1^2)=0.$$

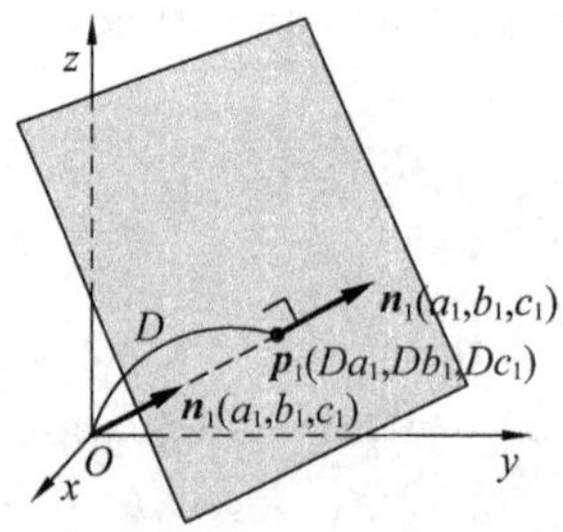

Figure 6.16 Plane with unit normal.

Since $\boldsymbol{n}$ is of length 1, this leads to

$$a_1x+b_1y+c_1z=D.$$

■ **Example 6.11** Find the distance from the origin for the plane of equation $x+2y+2z-6=0$.

The components of a normal vector are obtained by taking the coefficients of x, y, z from the equation. That is, $\boldsymbol{n}=(1,2,2)$. Since $\|\boldsymbol{n}\|=3$, we normalize $\boldsymbol{n}$ by dividing it to its norm to obtain the normalized vector $\left(\frac{1}{3}, \frac{2}{3}, \frac{2}{3}\right)$. Thus we divide both sides of the plane equation by 3 and obtain the equation $\frac{1}{3}x+\frac{2}{3}y+\frac{2}{3}z-2=0$, that is

$$\frac{1}{3}x+\frac{2}{3}+\frac{2}{3}z=2.$$

The distance from the origin to the plane is then equal to 2.

■

6.6.4 The distance from an arbitrary point to a given plane

The shortest distance from an arbitrary point $P_2(x_2, y_2, z_2)$ of position vector $\boldsymbol{p}_2$ to a plane can be calculated by projecting the vector $\boldsymbol{p}_2-\boldsymbol{p}_1$ where $\boldsymbol{p}_1$ is the position vector of an arbitrary point P_1 in the plane, to the normal vector $\boldsymbol{n}$ of the plane.

As shown in Figure 6.17, the distance between the plane and the point P_2 is

$$D=\|\boldsymbol{p}_2-\boldsymbol{p}_1\|\cos\theta=\frac{(\boldsymbol{p}_2-\boldsymbol{p}_1)\cdot\boldsymbol{n}}{\|\boldsymbol{n}\|}.$$

Assuming that the equation of the plane is $ax+by+cz-d=0$, a normal vector is $\boldsymbol{n}=(a,b,c)$ and consequently

$$D=\frac{|ax_2+by_2+cz_2-(ax_1+by_1+cz_1)|}{\sqrt{a^2+b^2+c^2}}.$$

Since P_1 belongs to the plane, it follows that $ax_1+by_1+cz_1=d$, which implies that

$$D=\frac{|ax_2+by_2+cz_2-d|}{\sqrt{a^2+b^2+c^2}}.$$

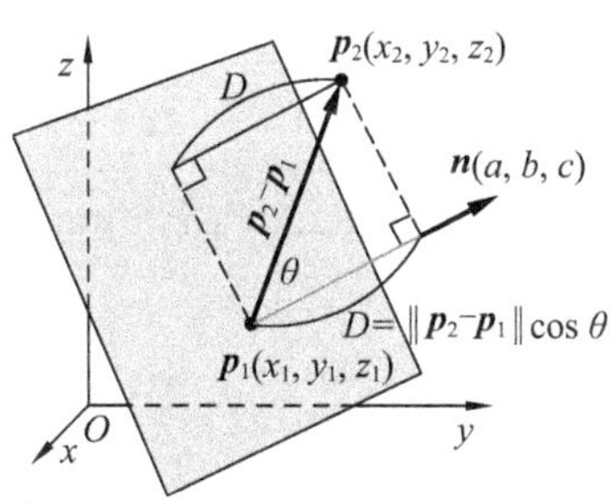

Figure 6.17 Distance between plane and point.

Ⓡ Notice the mnemonics: at the numerator, the coordinates of the point we compute the distance from are substituted into the equation of the plane, a modulus being taken since a distance should always be positive. The denominator contains the sum of coefficients of x, y, z, squared.

■ **Example 6.12** The distance from a point $(-1, -2, -3)$ to a plane $x+2y+2z-6=0$ is

$$D=\frac{|1(-1)+2(-2)+2(-3)-6|}{\sqrt{1^2+2^2+2^2}}$$
$$=\frac{17}{3}.$$

■

6.7 Summary

In mathematics, a **scalar** refers to a quantity described by a single real number, used to measure magnitude (size). A quantity which requires both a magnitude and a direction to be described completely is called a **vector.**

A vector is frequently represented by a directed line segment, connecting an **initial point** (or **tail**) with a **terminal point** (or **head**). The vector with initial point A and terminal point B is denoted by $\overrightarrow{AB}$.

Magnitude and direction

The **magnitude** also called **length** or **Euclidean norm** of $\overrightarrow{AB}$, denoted by $\| \overrightarrow{AB} \|$, is defined to be the length of the segment $\overrightarrow{AB}$, The **direction** of $\overrightarrow{AB}$ is defined to be the direction of the line segment AB.

Equality of vectors

Two vectors $\overrightarrow{AB}$ and $\overrightarrow{CD}$ are said to be **equal** or **equivalent** if they have the same magnitude and direction.

The null vector

If A coincides with B, one obtains the **null vector**, or **zero vector**, also denoted as 0, which has length 0 and whose direction is undefined.

Unit vector

A vector of length 1 is called an **unit vector.** Dividing any nonzero vector by its length, one obtains an unit vector.

Notation

If its initial and terminal point are not specified, a vector is usually denoted in lowercase boldface, as in $\boldsymbol{a}$, $\boldsymbol{b}$, $\boldsymbol{c}$, $\boldsymbol{u}$, $\boldsymbol{v}$, or with non-bold lowercase accented with a small arrow, as in $\vec{a}$, $\vec{b}$, $\vec{c}$, $\vec{u}$, $\vec{v}$.

The negative vector

Given a vector $\boldsymbol{a}$, a vector which has the same length as $\boldsymbol{a}$, but opposite direction, is denoted by $-\boldsymbol{a}$. The negative of a vector $\overrightarrow{AB}$ is the vector $\overrightarrow{BA}$. That is, $\overrightarrow{BA}=-\overrightarrow{AB}$.

Multiplication of a vector by a scalar

If c is a real number and $\boldsymbol{v}$ is a vector, then $c\boldsymbol{v}$ is a vector of magnitude $|c|\|\boldsymbol{v}\|$ which has the same direction as $\boldsymbol{v}$ if $c>0$ and opposite direction if $c<0$.

Parallel (collinear) vectors

Two vector $\boldsymbol{a}$ and $\boldsymbol{b}$ are parallel or collinear if they have the same direction. In this case, one of them is obtained from the other via multiplication with a constant.

Addition of vectors

Triangle law

To add vectors, one can use the **triangle law** (or **head-to-tail law**). Given two vectors $\boldsymbol{a}$ and $\boldsymbol{b}$, their sum $\boldsymbol{a}+\boldsymbol{b}$ may be computed by placing the tail of the vector $\boldsymbol{b}$ at the head of the vector $\boldsymbol{a}$, and then drawing a vector from the tail of $\boldsymbol{a}$ to the head of $\boldsymbol{b}$. The newly drawn vector represents the vector $\boldsymbol{a}+\boldsymbol{b}$.

One does not always need diagrams if letter correspondence is observed. For instance,

$$\overrightarrow{AB}+\overrightarrow{BC}=\overrightarrow{AC},$$

since the tail of the second vector, B, coincides with the head of the first vector. Then the sum vector, $\overrightarrow{AC}$, starts in the first point, A, and ends in the last point, C.

Also, for vectors, the straight way ($\overrightarrow{AC}$, directly from A to C) equals the deviating way ($\overrightarrow{AB}+\overrightarrow{BC}$, deviating through B).

Parallelogram law

To add vectors, one may also use the **parallelogram law**. Given two vectors $\boldsymbol{a}$ and $\boldsymbol{b}$, place the tail of $\boldsymbol{b}$ at the tail of $\boldsymbol{a}$ and draw the parallelogram determined by the sides $\boldsymbol{a}$ and $\boldsymbol{b}$. Then the sum $\boldsymbol{a}+\boldsymbol{b}$ is the diagonal of the parallelogram starting from the common tail.

Polygon law

To add more than two vectors, one can use the **polygon law**, which consists in the multiple use of the **triangle law**, by placing repeatedly the tail of the next vector at the head of the previous vector and then drawing a vector from the tail of the first vector to the head of the last vector. The newly drawn vector, which closes the polygon, represents the sum of all previous vectors.

One does not need diagrams if letter correspondence is observed. For instance,

$$\overrightarrow{AB}+\overrightarrow{BC}+\overrightarrow{CD}=\overrightarrow{AD}$$

by the repeated application of the triangle rule.

Again, the straight way ($\overrightarrow{AD}$, directly from A to D) equals the deviating way ($\overrightarrow{AB}+\overrightarrow{BC}+\overrightarrow{CD}$, deviating through both B and C). One may deviate through as many points as necessary.

If one adds together a set of vectors which return to the starting point, while the tail of the next vector is always at the head of the previous vector, then the result is the null vector. For instance, $\overrightarrow{AB}+\overrightarrow{BC}+\overrightarrow{CA}=\mathbf{0}$.

Subtraction of vectors

To find $\boldsymbol{a}-\boldsymbol{b}$, one simply adds $\boldsymbol{a}$ with the negative of $\boldsymbol{b}$. If $\boldsymbol{a}$ and $\boldsymbol{b}$ have the same tail, then $\boldsymbol{a}-\boldsymbol{b}$ is the vector drawn from the head of $\boldsymbol{b}$ to the head of $\boldsymbol{a}$.

Standard basis

We shall call the **standard basis** (**natural basis**) the set of unit vectors $\{\boldsymbol{i},\boldsymbol{j},\boldsymbol{k}\}$ which have the direction of the axes of a Cartesian coordinate system

The components of a vector

For any vector $\boldsymbol{a}$, there exist the numbers a_1, a_2, a_3, which are uniquely determined, such that

$$\boldsymbol{a}=a_1\boldsymbol{i}+a_2\boldsymbol{j}+a_2\boldsymbol{k}.$$

The numbers a_1, a_2, a_3 are called the **components**(or **coordinates**) of $\boldsymbol{a}$ with respect to the standard basis and, in this context, the vector $\boldsymbol{a}$ may also represented by indicating its components, in the form

$$\boldsymbol{a}=(a_1, a_2, a_3).$$

Equality of two vectors

Two vectors $\boldsymbol{a}$ and $\boldsymbol{b}$ are equal if and only if they have the same components.

The magnitude of a vector

If $\boldsymbol{a}=a_1\boldsymbol{i}+a_2\boldsymbol{j}+a_3\boldsymbol{k}$, then the magnitude of $\boldsymbol{a}$ is

$$\|\boldsymbol{a}\|=\sqrt{a_1^2+a_2^2+a_3^2}.$$

The null vector

The null vector is a vector with zero components, that is

$$\mathbf{0}=0\boldsymbol{i}+0\boldsymbol{j}+0\boldsymbol{k} \quad \text{or} \quad \mathbf{0}=(0,0,0).$$

The null vector has zero magnitude.

The negative or inverse of a vector

If $\boldsymbol{a}=a_1\boldsymbol{i}+a_2\boldsymbol{j}+a_3\boldsymbol{k}=(a_1, a_2, a_3)$, then its negative is

$$-\boldsymbol{a}=-a_1\boldsymbol{i}-a_2\boldsymbol{j}-a_3\boldsymbol{k} \text{ or } -\boldsymbol{a}=(-a_1,-a_2,-a_3).$$

Addition and subtraction of vectors

The addition and subtraction of vectors can be performed using the respective operations on their components. If $\boldsymbol{a}=a_1\boldsymbol{i}+a_2\boldsymbol{j}+a_3\boldsymbol{k}$ and $\boldsymbol{b}=b_1\boldsymbol{i}+b_2\boldsymbol{j}+b_3\boldsymbol{k}$, then

$$\boldsymbol{a}+\boldsymbol{b}=(a_1+b_1)\boldsymbol{i}+(a_2+b_2)\boldsymbol{j}+(a_3+b_3)\boldsymbol{k};$$
$$\boldsymbol{a}-\boldsymbol{b}=(a_1-b_1)\boldsymbol{i}+(a_2-b_2)\boldsymbol{j}+(a_3-b_3)\boldsymbol{k}.$$

Scalar multiplication

The scalar multiplication of a vector can be computed multiplying each component with that constant. If $\boldsymbol{a}=a_1\boldsymbol{i}+a_2\boldsymbol{j}+a_3\boldsymbol{k}$ and c is a constant, then

$$c\,\boldsymbol{a}=ca_1\boldsymbol{i}+ca_2\boldsymbol{j}+ca_3\boldsymbol{k}.$$

Parallel (collinear) vectors

Two vectors $\boldsymbol{a}=a_1\boldsymbol{i}+a_2\boldsymbol{j}+a_3\boldsymbol{k}$ and $\boldsymbol{b}=b_1\boldsymbol{i}+b_2\boldsymbol{j}+b_3\boldsymbol{k}$ are parallel if

$$\frac{a_1}{b_1}=\frac{a_2}{b_2}=\frac{a_3}{b_3}.$$

A denominator can be 0, with the convention that the corresponding numerator is also 0.

The position vector

To every point $P(a,b,c)$ one can uniquely associate its **position vector**

$$\overrightarrow{OP}=ai+bj+ck,$$

multiplying the components of P with the vectors of the standard basis and adding up the results.

The vector between two points

Let $A(x_1,y_1,z_1)$ and $B(x_2,y_2,z_2)$ be two points. Then

$$\begin{aligned}\overrightarrow{AB}&=\overrightarrow{OB}-\overrightarrow{OA}\\&=(x_2\boldsymbol{i}+y_2\boldsymbol{j}+z_2\boldsymbol{k})-(x_1\boldsymbol{i}+y_1\boldsymbol{j}+z_1\boldsymbol{k})\\&=(x_2-x_1)\boldsymbol{i}+(y_2-y_1)\boldsymbol{j}+(z_2-z_1)\boldsymbol{k}.\end{aligned}$$

Vectors in two dimensions

It is entirely possible to state similar algebraic considerations for vectors in a plane, rather than for vectors in space.

Standard basis

In this case, the standard basis will be $\{\boldsymbol{i},\ \boldsymbol{j}\}$.

The components of a vector

Any vector $\boldsymbol{a}$ has two components a_1, a_2, which are uniquely determined, such that

$$\boldsymbol{a}=a_1\boldsymbol{i}+a_2\boldsymbol{j},$$

and, in this context, the vector $\boldsymbol{a}$ may also represented by indicating its components, in the form

$$\boldsymbol{a}=(a_1,a_2).$$

The magnitude of a vector

If $\boldsymbol{a}=a_1\boldsymbol{i}+a_2\boldsymbol{j}$, then the magnitude (length) of $\boldsymbol{v}$ is

$$\|\boldsymbol{a}\|=\sqrt{a_1^2+a_2^2}.$$

For equality of vectors, collinearity of vectors, vector addition and subtraction and the multiplication of a vector by a scalar, one proceeds as in the three-dimensional case.

The scalar product

The scalar or dot product of two non-zero vectors $\boldsymbol{a}$ and $\boldsymbol{b}$ is defined as

$$\boldsymbol{a}\cdot\boldsymbol{b}=\|\boldsymbol{a}\|\|\boldsymbol{b}\|\cos\theta,$$

where θ is the angle between $\boldsymbol{a}$ and $\boldsymbol{b}$. If any of $\boldsymbol{a}$ and $\boldsymbol{b}$ are $\boldsymbol{0}$, the scalar product is defined as being 0.

Algebraic expression

If $\boldsymbol{a}=(x_1, x_2, x_3)$ and $\boldsymbol{b}=(y_1, y_2, y_3)$, then in terms of components the dot product of $\boldsymbol{a}$ and $\boldsymbol{b}$ is

$$\boldsymbol{a}\cdot\boldsymbol{b}=x_1y_1+x_2y_2+x_3y_3.$$

Similarly, if $\boldsymbol{a}=(x_1, x_2)$ and $\boldsymbol{b}=(y_1, y_2)$, then

$$\boldsymbol{a}\cdot\boldsymbol{b}=x_1y_1+x_2y_2.$$

Properties of the dot product

(1) Commutative: $\boldsymbol{a}\cdot\boldsymbol{b}=\boldsymbol{b}\cdot\boldsymbol{a}$.

(2) Distributive over vector addition: $\boldsymbol{a}\cdot(\boldsymbol{b}+\boldsymbol{c})=\boldsymbol{a}\cdot\boldsymbol{b}+\boldsymbol{a}\cdot\boldsymbol{c}$.

(3) Bilinear: $\boldsymbol{a}\cdot(r\boldsymbol{b}+\boldsymbol{c})=r(\boldsymbol{a}\cdot\boldsymbol{b})+(\boldsymbol{a}\cdot\boldsymbol{c})$.

(4) Scalar multiplication: $(c_1\boldsymbol{a})(c_2\boldsymbol{b})=c_1c_2(\boldsymbol{a}\cdot\boldsymbol{b})$.

(5) Not associative because the dot product between a scalar $(\boldsymbol{a}\cdot\boldsymbol{b})$ and a vector $(\boldsymbol{c})$ is not defined.

(6) Orthogonal: Two non-zero vectors $\boldsymbol{a}$ and $\boldsymbol{b}$ are orthogonal if and only if $\boldsymbol{a}\cdot\boldsymbol{b}=0$.

(7) No cancellation: Unlike multiplication of ordinary numbers, where if $ab=ac$, then b always equals c unless a is zero, the dot product does not obey the cancellation law.

The angle between two vectors

$$\cos\theta=\frac{\|\boldsymbol{a}\cdot\boldsymbol{b}\|}{\|\boldsymbol{a}\|\|\boldsymbol{b}\|},$$

where θ is the angle between the vectors $\boldsymbol{a}$ and $\boldsymbol{b}$.

The vector product

The vector or cross product of two vectors $\boldsymbol{a}$ and $\boldsymbol{b}$ is defined only in the three-dimensional space and is given by

$$\boldsymbol{a}\times\boldsymbol{b}=\|\boldsymbol{a}\|\|\boldsymbol{b}\|\sin\theta\boldsymbol{n},$$

where θ is the angle between $\boldsymbol{a}$ and $\boldsymbol{b}$ in the plane containing them and $\boldsymbol{n}$ is a unit vector perpendicular to the plane containing $\boldsymbol{a}$ and $\boldsymbol{b}$ in the direction given by the right-hand rule.

The cross-product is anti-commutative, i.e.,

$$\boldsymbol{b}\times\boldsymbol{a}=-(\boldsymbol{a}\times\boldsymbol{b}).$$

Algebraic expression

If

$$\boldsymbol{u}=u_1\boldsymbol{i}+u_2\boldsymbol{j}+u_3\boldsymbol{k} \text{ and } \boldsymbol{v}=v_1\boldsymbol{i}+v_2\boldsymbol{j}+v_3\boldsymbol{k},$$

then

$$\begin{aligned}\boldsymbol{u}\times\boldsymbol{v}&=(u_1\boldsymbol{i}+u_2\boldsymbol{j}+u_3\boldsymbol{k})\times(v_1\boldsymbol{i}+v_2\boldsymbol{j}+v_3\boldsymbol{k})\\&=u_1v_1(\boldsymbol{i}\times\boldsymbol{i})+u_1v_2(\boldsymbol{i}\times\boldsymbol{j})+u_1v_3(\boldsymbol{i}\times\boldsymbol{k})+u_2v_1(\boldsymbol{j}\times\boldsymbol{i})+u_2v_2(\boldsymbol{j}\times\boldsymbol{j})+\\&\quad u_2v_3(\boldsymbol{j}\times\boldsymbol{k})+u_3v_1(\boldsymbol{k}\times\boldsymbol{i})+u_3v_2(\boldsymbol{k}\times\boldsymbol{j})+u_3v_3(\boldsymbol{k}\times\boldsymbol{k})\\&=(u_2v_3-u_3v_2)\boldsymbol{i}+(u_3v_1-u_1v_3)\boldsymbol{j}+(u_1v_2-u_2v_1)\boldsymbol{k}.\end{aligned}$$

The area of a parallelogram

The magnitude of the cross product $\boldsymbol{a}\times\boldsymbol{b}$ can be interpreted as the positive area of the parallelogram having $\boldsymbol{a}$ and $\boldsymbol{b}$ as sides,

$$\|\boldsymbol{a}\times\boldsymbol{b}\|=\|\boldsymbol{a}\|\|\boldsymbol{b}\|\sin\theta.$$

The volume of a parallelepiped

The volume V of a parallelepiped having $\boldsymbol{a}$, $\boldsymbol{b}$ and $\boldsymbol{c}$ as edges is given by

$$V=\|\boldsymbol{a}\cdot(\boldsymbol{b}\times\boldsymbol{c})\|.$$

Direction cosines

Suppose we have a vector $\overrightarrow{OA}$ with initial point at the origin and terminal point at A. Suppose also that we have a unit vector in the same direction as $\overrightarrow{OA}$ and let the unit vector be

$$\boldsymbol{u}=\mu_1\boldsymbol{i}+\mu_2\boldsymbol{j}+\mu_3\boldsymbol{k}.$$

If α is the angle between $\boldsymbol{u}$ and the semi-axis Ox, β is the angle between $\boldsymbol{u}$ and the semi-axis Oy and γ is the angle between $\boldsymbol{u}$ and the semi-axis Oz, then

$$\boldsymbol{u}=\cos\alpha\boldsymbol{i}+\cos\beta\boldsymbol{j}+\cos\gamma\boldsymbol{k}.$$

These 3 cosines are called the **direction cosines.**

Equations of a straight line

The vector equation of a straight line

The straight line through the point A(with position vector $\boldsymbol{a}$) parallel to the vector $\boldsymbol{d}$ has the equation

$$\boldsymbol{r}=\boldsymbol{a}+\lambda\boldsymbol{d}.$$

The standard Cartesian equation of a straight line

The equation of the straight line passing through $P_1(x_1, y_1, z_1)$ and parallel to $\boldsymbol{d}=d_1\boldsymbol{i}+d_2\boldsymbol{j}+d_3\boldsymbol{j}$ is

$$\frac{x-x_1}{d_1}=\frac{y-y_1}{d_2}=\frac{z-z_1}{d_3}.$$

Equations of a plane

The vector equation of a plane

Assume that we know a point that is on the plane, $P_1(x_1, y_1, z_1)$ with position vector $\boldsymbol{p}_1$. Let us also suppose that there is a vector that is orthogonal (perpendicular) to the plane, $\boldsymbol{n}=a\boldsymbol{i}+b\boldsymbol{j}+c\boldsymbol{k}$, called the **normal vector.** Assume that $P(x, y, z)$ is an arbitrary point on the plane, with position vector $\boldsymbol{p}$. Then **the vector equation of a plane** is

$$(\boldsymbol{p}-\boldsymbol{p}_1)\cdot\boldsymbol{n}=0.$$

The standard Cartesian equation of a plane

If $P_1(x_1, y_1, z_1)$ is a point on the plane and $\boldsymbol{n}=a\boldsymbol{i}+b\boldsymbol{j}+c\boldsymbol{k}$, is the normal vector, then, if we denote $d=(ax_1+by_1+cz_1)$, the equation of the plane is

$$ax+by+cz=d.$$

The distance from the origin to a given plane

If the normal vector is taken to be of length 1 (normalized), then the corresponding constant term of the plane equation is the distance from the origin to the plane.

The distance from an arbitrary point to a given plane

The shortest distance from an arbitrary point $P_2(x_2, y_2, z_2)$ to a plane of equation $ax+by+cz-d=0$ is

$$D=\frac{|ax_2+by_2+cz_2-d|}{\sqrt{a^2+b^2+c^2}}.$$

At the numerator, the coordinates of the point we compute the distance from are substituted into the equation of the plane, a modulus being taken since a distance should always be positive. The denominator contains the sum of coefficients of x, y, z, squared.

6.8 Further problems

Problem 6.1 Given $\boldsymbol{a}=(2,1)$, $\boldsymbol{b}=(3,4)$ and $\boldsymbol{c}=(2,5)$, evaluate

(1) $3\boldsymbol{a}-4\boldsymbol{c}$;

(2) $2\boldsymbol{b}+3\boldsymbol{c}$.

Problem 6.2 If $\overrightarrow{PQ}=(2, 4)$ and $\overrightarrow{PR}=(3,-1)$, find $\overrightarrow{QR}$.

Problem 6.3 If $\overrightarrow{PQ}=(8,-9)$ and $\overrightarrow{PR}=(2,5)$, find $\|\overrightarrow{QR}\|$.

Problem 6.4 If $\boldsymbol{p}=(3,4)$ and $\boldsymbol{q}=(-2,7)$, find

(1) $4\boldsymbol{p}-2\boldsymbol{q}$;

(2) $|4\boldsymbol{p}-2\boldsymbol{q}|$.

Problem 6.5 If $\boldsymbol{p}=(-4, 6)$, $\boldsymbol{q}=(10,-5)$ and $k\boldsymbol{p}-l\boldsymbol{q}=(-2,3)$, find the values of

k and l.

Problem 6.6 Given $\boldsymbol{a}=(1,2,-1)$, $\boldsymbol{b}=(2, 1,0)$ and $\boldsymbol{c}=(0,-2,-1)$, evaluate

(1) $\boldsymbol{a}+\boldsymbol{b}$;

(2) $\boldsymbol{b}-\boldsymbol{c}$;

(3) $\boldsymbol{a}+\boldsymbol{b}+\boldsymbol{c}$;

(4) $3\boldsymbol{a}-\boldsymbol{b}-2\boldsymbol{c}$.

Problem 6.7 Let $ABCD$ be a quadrilateral and let A, B and D have coordinates $(0,2)$, $(2,5)$ and $(8,0)$ respectively. If $\overrightarrow{AD}=2\overrightarrow{BC}$, find the coordinates of C.

Problem 6.8 Let $O(0,0)$, $A(3,1)$ and $B(2,-1)$ be three vertices of a quadrilateral $OABC$. If $\overrightarrow{BC}=(3,-2)$, find

(1) the coordinates of C;

(2) $\overrightarrow{AC}$;

(3) $\|\overrightarrow{OC}\|$.

Problem 6.9 Let A, B and C be points in the Cartesian plane. If the coordinates of A are $(3,-2)$, $\overrightarrow{AB}(4,-3)$ and $\overrightarrow{AC}=(4,9)$, calculate

(1) the coordinates of B and C;

(2) the magnitude of $\overrightarrow{BC}$.

Problem 6.10 If $\boldsymbol{a}=(5,4,1)$ and $\boldsymbol{b}=(-3,4,2)$, find $|\boldsymbol{a}-\boldsymbol{b}|$.

Problem 6.11 The points $A(-2,1)$, $B(2,1)$, $C(1,4)$ and $D(p, q)$ are the vertices of a parallelogram $ABCD$.

(1) Write down $\overrightarrow{AB}$, $\overrightarrow{DC}$, and $\overrightarrow{BD}$ as column vectors and deduce the values of p and q.

(2) Calculate the area of the parallelogram $ABCD$.

Problem 6.12 Given the points $P(-2,3)$, $Q(10,4)$, $R(3,12)$ and $S(4,0)$, show that $\overrightarrow{PQ}$ is perpendicular to $\overrightarrow{RS}$.

Problem 6.13 If the vectors $\boldsymbol{a}=x\boldsymbol{i}+y\boldsymbol{j}+z\boldsymbol{k}$ and $\boldsymbol{b}=3\boldsymbol{i}-\boldsymbol{j}+2\boldsymbol{k}$ are perpendicular, find y when $x=2$ and $z=1$.

Problem 6.14 Using the scalar product, find the angle between the vectors $\boldsymbol{p}=(-\sqrt{3},1)$ and $\boldsymbol{q}=(0,-3)$.

Problem 6.15 Calculate the area of the parallelogram spanned by the vectors

(1) $\boldsymbol{a}=(3,-3,1)$ and $\boldsymbol{b}=(4,9,2)$;

(2) $\boldsymbol{a}=(3,-3,1)$ and $\boldsymbol{b}=(-12,12,-4)$.

Problem 6.16 Given the points $A(2,4)$ and $B(12, 8)$.

(1) Find the distance between these two points.

(2) Find the midpoint between these two points.

Problem 6.17 The points $R(-2,4)$, $Q(9, 6)$ and $P(4,-4)$ and S are the vertices of a parallelogram $RQPS$ and M is the midpoint of PR.

(1) Find the coordinates of M and S.

(2) Show that the parallelogram $RQPS$ is a rhombus.

Answers

Problem 6.1 (1) $(-2,-17)$. (2) $(12,23)$.

Problem 6.2 $\overrightarrow{QR}=(1,-5)$.

Problem 6.3 $\|\overrightarrow{QR}\|=2\sqrt{58}$.

Problem 6.4 (1) $(16,2)$. (2) $2\sqrt{65}$.

Problem 6.5 $k=\frac{1}{2}$ and $l=0$.

Problem 6.6 (1) $(3,3,-1)$. (2) $(2,3,1)$. (3) $(3,1,-2)$. (4) $(1,9,-1)$.

Problem 6.7 $(6,4)$.

Problem 6.8 (1) $(5,-3)$. (2) $(2,-4)$. (3) $\sqrt{34}$.

Problem 6.9 (1) $B(-1,1)$, $C(7,7)$. (2) 10.

Problem 6.10 $\sqrt{65}$.

Problem 6.11 (1) $p=-3$, $q=4$. (2) 12.

Problem 6.12 Since $\overrightarrow{PQ}=(12,1)$ and $\overrightarrow{RS}=(1,-12)$, $\overrightarrow{PQ}\cdot\overrightarrow{RS}=0$, which implies that $\overrightarrow{PQ}$ is perpendicular to $\overrightarrow{RS}$.

Problem 6.13 $y=8$.

Problem 6.14 $\frac{\pi}{3}$.

Problem 6.15 (1) $5\sqrt{70}$. (2) 0.

Problem 6.16 (1) $2\sqrt{29}$. (2) $(7,6)$.

Problem 6.17 (1) $M(1,0)$ and $S(-7,6)$. (2) Since $\|\overrightarrow{PQ}\|=\|\overrightarrow{QR}\|=5\sqrt{5}$, the parallelogram $RQPS$ is a rhombus.

Chapter 7 Mathematical arguments and proofs

7.1 Mathematical logic

Mathematical logic, used to construct mathematical arguments, operates primarily with statements and with their truth values. A **statement** is a proposition that can be determined to be true or false. The **truth value** of a statement is then either "true", shortened as T, or "false", shortened as F.

Logical operations

In logic, there are certain operations for combining or modifying statements; some of these operations are "and", "or", "not", and "if ⋯ then". In mathematics, these words have precise meanings and notations, which are given below.

Definition 7.1.1—Not. The simplest logical operation is "not". If P is a statement, then the statement "not P" is called the negation of P. When P is true, "not P" is false and when P is false, "not P" is true. The **truth table** of this operation is given below:

P	$\neg P$
T	F
F	T

Notation 7.1 The statement "not P" is denoted by $\neg P$.

Definition 7.1.2—And. If P and Q are two statements, then the statement "P and Q" is defined to be true, provided that P and Q are both true. When P is false or Q is false or both P and Q are false, we say that "P and Q" is false, as indicated below:

P	Q	$P \wedge Q$
T	T	T
T	F	F
F	T	F
F	F	F

Notation 7.2 The statement "P and Q" is denoted by $P \wedge Q$.

Definition 7.1.3—Or. If P and Q are two statements, then the statement "P or Q" is defined to be false, provided that P and Q are both false. When P is true or Q is true or both P and Q are true, "P or Q" is defined to be true, as seen in the next truth table:

P	Q	$P \vee Q$
T	T	T
T	F	T
F	T	T
F	F	F

Notation 7.3 The statement "P or Q" is denoted by $P \vee Q$.

Definition 7.1.4—If then. If P and Q are two statements, then the statement "if P then Q" is defined to be true when P and Q are both true or both false, or P is false. When P is true and Q is false, "if P then Q" is false, as seen below:

P	Q	$P \Rightarrow Q$
T	T	T
T	F	F
F	T	T
F	F	T

Notation 7.4 We shall commonly abbreviate the statement "if P then Q" by "P implies Q" or "$P \Rightarrow Q$",

Definition 7.1.5—If and only if. If P and Q are two statements, then the statement "P if and only if Q" is defined to be true, when P and Q are both true or both false. When one of P, Q is true and the other is false, "P if and only if Q" is false, as indicated below:

P	Q	$P \Leftrightarrow Q$
T	T	T
T	F	F
F	T	F
F	F	T

Notation 7.5 We shall commonly abbreviate the statement "P if and only if Q" by "P iff Q" or "$P \Leftrightarrow Q$". Also, when $P \Leftrightarrow Q$ is true, we say P and Q are **equivalent**.

Quantifiers

The phrases "for all" and "there exists" are called **quantifiers**.

Definition 7.1.6—Universal and existential quantifiers. The quantifier "for all" is called the **universal** quantifier, while the quantifier "there exists" is called the **existential** quantifier.

Notation 7.6 "$\forall$" reads "for all" (or "for every", or yet "for each") and "$\exists$" reads "there exists".

As an example for the use of quantifiers, we can give precise definitions of the terms "even" and "odd".

■ **Example 7.1** An integer x is even if there exists an integer z such that $x=2z$. (Here the "if" in this definition is really an "if and only if". Mathematical literature tends to misuse the word "if" this way when making definitions, and we will do this as well). An integer x is odd if there exists an integer z such that $x=2z+1$.

■

Order of quantifiers

The order of quantifiers is very important. Changing the order of the quantifiers in a statement will often change the meaning (and truth value) of a statement. For example, the statement

$$(\forall x\in \mathbf{Z})(\exists y\in \mathbf{Z})x<y,$$

which reads as "for every integer x there is an integer y such that x is smaller than y" is true. However, the statement

$$(\exists y\in \mathbf{Z})(\forall x\in \mathbf{Z})x<y,$$

which reads as "there is an integer y such that for every integer x, x is smaller than y", is false.

Negation of quantifiers

It is also important (for instance, for the method of Proof by Contradiction mentioned below) to know how to negate statements involving quantifiers. To negate a statement with quantifiers, we negate all the quantifiers first, from left to right (keeping the same order), process in which $\forall$ changes into $\exists$ and $\exists$ changes into $\forall$, then we take the negation of the statement. That is, the negation of "$(\forall x)P(x)$" is "$(\exists x)\neg P(x)$" and the negation of "$(\exists x)P(x)$" is "$(\forall x)\neg P(x)$". For instance, the negation of "All leaves are green" is "There is a leaf which is not green".

7.2 Proof by exhaustion

A mathematical proof is an argument which shows that a statement is always true by way of logical reasoning, rather than by enumerating several confirmatory examples.

Proof by exhaustion, also known as **proof of cases**, is a method of direct mathematical proof in which the statement to be proved is split into a finite number of cases or sets of equivalent cases and each of the cases is checked to see if the proposition in question holds. A proof by exhaustion includes two scenarios:

(1) Each situation or parameter value in the statement to be shown matches the conditions of (at least) one of the cases.

(2) Each of the cases is proven.

Ⓡ In theory, the proof by exhaustion method can be applied whenever the number of the cases is finite. Sometimes there are only two or three cases. Sometimes there may be thousands or even millions. However, mathematicians prefer to avoid proofs by exhaustion with large numbers of cases, which are viewed as inelegant. Also, since most mathematical sets are infinite, this method is seldom used to derive general mathematical results.

Exercise 7.1 Prove that the integer $x(x+1)$ is even for every integer x.

■

Proof Let x be any integer. Then x is either even or odd and there are consequently two cases to be analyzed. We will show that $x(x+1)$ is even in both of these cases.

Case 1 Suppose that x is even. Choose an integer k such that $x=2k$, which implies that $x(x+1)=2k(2k+1)$. Let $y=k(2k+1)$. Then y is an integer and $x(x+1)=2y$, so $x(x+1)$ is even.

Case 2 Suppose that x is odd. Choose an integer k such that $x=2k+1$. Then $x(x+1)=(2k+1)(2k+2)$. Let $y=(2k+1)(k+1)$. Then y is an integer and $x(x+1)=2y$, so $x(x+1)$ is even.

■

Exercise 7.2 Prove that every integer that is a perfect cube is a multiple of 9, or is 1 more than a multiple of 9, or is 1 less than a multiple of 9.

■

Proof Every integer n is either a multiple of 3, or 1 more than a multiple of 3, or 1 less than a multiple of 3. So these three cases are exhaustive:

Case 1 If $n=3p$, then $n^3=27p^3=9\cdot 3p^3$, which is a multiple of 9.

Case 2 If $n=3p+1$, then $n^3=27p^3+27p^2+9p+1=9(3p^3+3p^2+p)+1$, which is 1 more than a multiple of 9.

Case 3 If $n=3p-1$, then $n^3=27p^3-27p^2+9p-1=9(3p^3-3p^2+p)-1$, which is 1 less than a multiple of 9.

■

Exercise 7.3 Prove that every year in which the FIFA World Cup final tournament was held is divisible by 2, but not by 4.

■

Proof The statement can be proved by exhaustion (it's easier if one is a football fan) via listing out every year in which the FIFA World Cup final tournaments were held, and checking that every one of them is divisible by 2, but not by 4. With 20 total FIFA World Cup final tournaments as of 2014, this is a proof by exhaustion with 20 cases.

■

7.3 Proof by counterexample

Proof by counterexample is typically used for disproving something. Many times, a conjecture (a statement whose truth value is yet to be determined at the moment it is made) will make an overly presumptuous claim, stating that a certain property holds for all of something. To disprove it, one merely needs to find a single instance that satisfies the hypothesis, but not the conclusion. This type of argument allows us to prove that a property is not true by providing just a single example where it does not hold.

■ **Example 7.2** The great French mathematician Pierre de Fermat (1601—1665) conjectured that positive integers of the form

$$F_n=2^{2n}+1$$

(now known as Fermat numbers) are all prime. Certainly, the first five, namely, $F_0=3$, $F_1=5$, $F_2=17$, $F_3=257$ and $F_4=65\,537$ are all prime, but Leonhard Euler in 1752 spoilt the party by showing that the sixth Fermat number was composite. That is,

$$F_5=2^{2^5}+1=4\,294\,967\,297=641\times 6\,700\,417.$$

■

■ **Example 7.3** The great Swiss mathematician Leonhard Euler (1707—1783) once conjectured that it was impossible to find integers a, b, c, d and e that satisfy the equation

$$a^5+b^5+c^5+d^5=e^5.$$

In 1966, a counterexample was given, being observed that

$$27^5+84^5+110^5+133^5=144^5.$$

■

Exercise 7.4 Prove that not all triangles are obtuse.

■

Proof A counterexample is obvious: the equilateral triangle, having all angles equal to 60°. In this case, there are infinitely many counterexamples. However, it only takes one.

■

Exercise 7.5 Prove that not all graphs of linear functions are perpendicular to each other.

■

Proof To prove that this is true, we need to find a pair of linear functions whose graphs are not perpendicular. We suggest the following counterexample:

$$f_1(x)=3x+4,$$
$$f_2(x)=x-2.$$

The slope of the first graph (a line) is 3, while the slope of the second graph (another line) is 1. The product of slopes is 3, not -1, and therefore, the graphs are not perpendicular.

■

Ⓡ When proving by counterexample, it is not enough to state the counterexample. We must also explain why it is a counterexample.

7.4 Proof by contradiction

In logic, proof by contradiction is a form of indirect proof that establishes the truth or validity of a statement by assuming that it is false and deducing a contradiction. Suppose that we want to prove that the statement P is true. We begin by assuming that P is false. We then try to deduce a contradiction, i.e. some statement Q which we know is false. If we succeed, then our assumption that P is false must be wrong. So P is true, and our proof is completed.

Exercise 7.6 Prove that if x is rational and y is irrational, then $x+y$ is irrational. ■

Proof Let us assume the negation of what we are trying to prove: there exist a rational number x and an irrational number y such that $x+y$ is rational. One observes that

$$y=(x+y)-x.$$

Since $x+y$ and x are rational by assumption, and the difference of two rational numbers is rational, y is rational. But that contradicts our assumption, since y was supposed to be irrational. So our assumptions cannot be correct. So if x is rational and y is irrational, then $x+y$ is irrational.

■

Exercise 7.7 Prove that $\sqrt{2}$ is irrational. ■

Proof Suppose that $\sqrt{2}$ is rational. It follows from that the definition of rational numbers that $\sqrt{2}=\frac{a}{b}$ for some integers a and b, $b\neq 0$.

We can assume that b is positive, since otherwise we can simply change the signs of both a and b. Let this fraction be fully reduced. In particular, this means a and b are not both even. Squaring both sides of the equation $\sqrt{2}=\frac{a}{b}$ and multiplying both sides by b^2, we obtain $a^2=2b^2$. Since a^2 is even, it follows that a is even. Thus $a=2k$ for some integer k, so $a^2=4k^2$, and hence $b^2=2k^2$. Since b^2 is even, it follows that b is even. Consequently, a and b are both even, which is a contradiction. Therefore $\sqrt{2}$ cannot be rational.

■

Exercise 7.8 Prove that $a^2-4b\neq 2$ for any integers a, b. ■

Proof Suppose this statement is false. Thus there exist integers a and b such that

$$a^2-4b=2. \tag{7.1}$$

From this equation we obtain that $a^2=2(2b+1)$, so a^2 is even, which implies that a is even. Let then $a=2c$ for some integer c. Now plugging $a=2c$ back into the equation (7.1), we get that $4c^2-4b=2$. Dividing by 2, we obtain that $2(c^2-b)=1$, which is a contradiction (the left-hand side of the equality is an even number, while the right-hand side is odd).

■

7.5 Proof by induction

Mathematical induction is a form of direct proof of a statement involving a natural number, usually done in two steps. The first step, known as **the base case**, proves a given statement for a certain value of that natural number. The second step, known as **the inductive step**, proves that if the statement is assumed to be true for a natural number, then it must be true for the next natural number as well. Then the rule of inference establishes the statement to be true for all natural numbers larger than or at least equal to the base case.

Induction and dominoes

By analogy, mathematical induction can be informally illustrated by reference to the sequential effect of falling dominoes (see Figure 7.1). If we push over the first domino, and if each domino knocks over the next domino as it falls, then eventually every domino will fall. This reasoning is called the principle of mathematical induction. In fact, this principle can be regarded as one of the axioms defining the natural numbers. Let us state it precisely.

Figure 7.1 The dominoes falling in a line.

7.5.1 Principle of Mathematical Induction (PMI)

Assume that for each positive integer n we have a statement $P(n)$. If the following are true:

(1) (**base case**) $P(1)$ is true;

(2) (**inductive step**) for all $n \geqslant 1$, if $P(n)$ is true, then $P(n+1)$ is also true,

then $P(n)$ is true for all positive integers n.

The logic behind this principle is clear: by (1), the first statement $P(1)$ is true; By (2) with $n=1$, we obtain that $P(1) \Rightarrow P(2)$, hence $P(2)$ is true; By (2) again with $n=2$, we get that $P(2) \Rightarrow P(3)$, hence $P(3)$ is true; and so on.

Exercise 7.9 Prove that

$1+3+5+\cdots+2n-1=n^2$ for all positive integers n.

■

Proof Here, for n being a positive integer, $P(n)$ is the statement

$$1+3+5+\cdots+2n-1=n^2.$$

Base case: When $n=1$, we have $1=1$, and $P(1)$ is true.

Inductive step: Suppose that $P(n)$ is true. Then for a given $n \in \mathbf{Z}^+$,

$$1+3+5+\cdots+2n-1=n^2.$$

Adding $2n+1$ to both sides gives

$$1+3+5+\cdots+2n-1+2n+1=n^2+2n+1=(n+1)^2,$$

and consequently $P(n+1)$ is also true. Thus we have shown that $P(n) \Rightarrow P(n+1)$. We have now established parts (1) and (2). Hence by the Principle of Mathematical Induction, $P(n)$ is true for all positive integers n.

■

Exercise 7.10 Prove that, if $p>1$, then $(1+p)^n \geqslant 1+np$.

■

Proof Here, for n being a positive integer, $P(n)$ is the statement

$$P(n): \text{If } p>-1, \text{ then } (1+p)^n \geqslant 1+np.$$

For (1), it is obvious that $P(1)$ is true, as $1+p \geqslant 1+p$. For (2), suppose that the statement $P(n)$ is true, so $(1+p)^n \geqslant 1+np$. Since $p>-1$, one notes that $1+p>0$. We then multiply both sides of the inequality above by $1+p$ to obtain that

$$(1+p)^{n+1} \geqslant (1+np)(1+p)=1+(n+1)p+np^2 \geqslant 1+(n+1)p$$

and consequently $P(n+1)$ is also true. Thus we have shown that $P(n) \Rightarrow P(n+1)$. Hence, by induction, $P(n)$ is true for all positive integers n.

■

Essentially, induction is a basic tool for breaking down more complicated arguments into simple steps. There are many variants of the principle of induction. For example, one can start at 0 instead of 1, to prove that some statement is true for all nonnegative integers. Sometimes, one needs to start at another natural number, which is given away by the requirements of the statement to be proved. For instance, to prove that $2^n \leqslant n^2$ for every integer $n \geqslant 5$, one needs to start with $n=5$. One can also prove a statement about several positive integers $m, n, \cdots$ by doing induction on one variable at a time.

Under certain circumstances, it is easier to prove that $P(n+1)$ is true if one assumes that not only $P(n)$ is true, but also the other statements formulated up to that point are true as well, starting with the base case. Since this represents a "stronger foundation" for the inductive step, the resulting principle is called the Principle of Strong Mathematical Induction (although, in the end, the Principle of Strong Mathematical Induction and the Principle of Strong Mathematical Induction are equivalent).

7.5.2 Principle of Strong Mathematical Induction (PSMI)

Let k be a positive integer. Suppose that for each positive integer $n \geqslant k$ we have a statement $P(n)$. If we show the following two parts:

(1) $P(k)$ is true;

(2) For all $n \geqslant k$, if $P(m)$ is true for all integers m with $k \leqslant m \leqslant n$, then $P(n+1)$ is also true,

then $P(n)$ is true for all positive integers $n \geqslant k$.

Exercise 7.11 Prove that the sum of the internal angles in an $n-$ sided polygon is $(n-2)\pi$. ■

Proof Let $P(n)$ be the statement

$P(n)$: The sum of the internal angles in an n-sided polygon is $(n-2)\pi$.

Then $P(n)$ only makes sense if $n \geqslant 3$, while $P(1)$ and $P(2)$ make no sense because there is no 1-sided or 2-sided polygon.

For (1) we have $k=3$ in the above Principle, and $P(3)$ is true due to the sum of the angles in a triangle being $(3-2)\pi=\pi$.

For (2), as shown in Figure 7.2, consider an $(n+1)$-sided polygon with corners A_1, A_2, $\cdots$, A_{n+1}. Drawing the line A_1A_3, we then obtain that $A_1A_3A_4 \cdots A_{n+1}$ is an n-sided polygon. Since $P(n)$ is true, the sum of the internal angles in an n-sided polygon is $(n-2)\pi$. It follows from Figure 7.2 that the sum of the internal angles in an $(n+1)$-sided polygon $A_1A_2A_3 \cdots A_{n+1}$ is equal to the sum of those in the n-sided polygon $A_1A_3A_4 \cdots A_{n+1}$ plus the sum of those in the triangle $A_1A_2A_3$, that is,

$$(n-2)\pi+\pi=(n+1)-2\pi.$$

Hence, by strong induction, $P(n)$ is true for all $n \geqslant 3$.

■

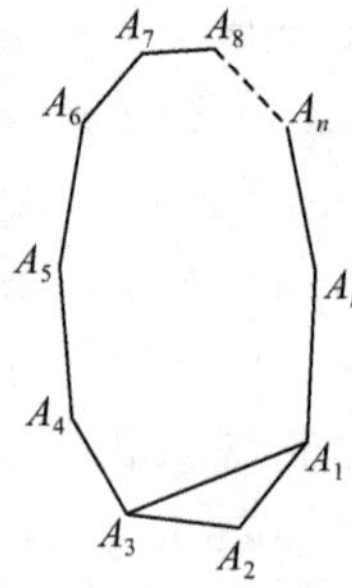

Figure 7.2 An $(n+1)$-sided polygon $A_1A_2A_3\cdots A_{n+1}$ $(n\geqslant 2)$.

Exercise 7.12 Prove that every positive integer greater than 1 is equal to a product of prime numbers.

■

Proof For $n\geqslant 2$, let $P(n)$ be the statement

$$P(n):n \text{ is equal to a product of prime numbers.}$$

For (1), it is seen that $P(2)$ is true. In what follows, for (2), suppose that $P(2),\cdots,P(n)$ are all true, which means that every number between 2 and n has a prime factorization. Now consider $P(n+1)$. If $n+1$ is prime, then it can be seen as a product of 1 prime. Otherwise, there is an integer a dividing $n+1$ such that $a\neq 1$ or $a\neq n+1$. Denoting $b=\frac{n+1}{a}$, we then have that

$$n+1=ab \text{ and } a,b\in\{2,3,\cdots,n\}.$$

It follows from the assumption that $P(a)$ and $P(b)$ are both true. That is, a and b have respective prime factorizations, i.e.,

$$a=p_1\ p_2\cdots p_n,\ b=q_1q_2\cdots q_n,$$

in which all the p_i and q_i are prime numbers. Hence

$$n+1=ab=p_1\ p_2\cdots p_nq_1q_2\cdots q_n$$

and $P(n+1)$ is true. We have now shown that $P(n)\Rightarrow P(n+1)$. Therefore, by strong induction, $P(n)$ is true for all $n\geqslant 2$.

■

7.6 Summary

Truth tables

P	Q	$P\vee Q$	$P\wedge Q$	$P\Rightarrow Q$	$P\Rightarrow Q$
T	T	T	T	T	T
T	F	T	F	F	F
F	T	T	F	T	F
F	F	F	F	T	T

The first 2 columns are P and Q individual states (either true (T) or false (F)); The third column is "P or Q", with the "$\vee$" symbolizing "or"; The fourth column is "P and Q", with the "$\wedge$" symbolizing "and"; The fifth column is "P implies Q", with the "$\Rightarrow$" symbolizing "implies"; The sixth column is "Q if and only if Q", with the "$\Leftrightarrow$" symbolizing "if and only if".

Proof by exhaustion (Proof by cases)

Proof by exhaustion is a method of direct mathematical proof in which the statement to be proved is split into a finite number of cases or sets of equivalent cases and each of the cases is checked to see if the proposition in question holds.

Proof by counterexample

Proof by counterexample is to find one instance that satisfies the hypothesis, but not the conclusion, and then to disprove the initial statement.

Proof by contradiction

Proof by contradiction is to assume that the statement we want to prove is false, and then show that this assumption leads to nonsense. We are then led to conclude that we were wrong to assume the statement was false, so the statement must be true.

Proof by induction

Proof by induction, including Principle of Mathematical Induction (**PMI**) and Principle of Strong Mathematical Induction (**PSMI**), is a form of direct proof, usually done in two steps. When trying to prove a given statement formulated for a set of natural numbers, the following steps must be taken.

(1) The first step, known as the base case, is to prove the given statement for the first natural number in the set.

(2) The second step, known as the inductive step, is to prove that, if the statement is assumed to be true for a natural number (for **PMI**), or for all natural numbers from the first natural number in the set up to that natural number (for **PSMI**), then it must be true for the next natural number as well.

Having proved these two steps, the rule of inference establishes the statement to be true for all natural numbers in the set.

7.7 Further problems

Problem 7.1 If P and Q are two statements, write the truth tables of the statements $(\neg P)\vee Q$, $P\Rightarrow(\neg Q)$, $P\wedge(\neg Q)$.

Problem 7.2 If P and Q are two statements, write the truth tables of the statements $P\Rightarrow(P\vee Q)$, $(P\wedge Q)\Rightarrow P$. What do you observe?

Problem 7.3 If P and Q are two statements, prove that the statements $P \vee Q$ and $(\neg P) \wedge (\neg Q)$ are equivalent.

Problem 7.4 Prove that for every integers x and y, if x is odd and y is odd then xy is odd.

Problem 7.5 Prove that for every integer x, $x+4$ is odd if and only if $x+7$ is even.

Problem 7.6 Prove that for every odd integer x there exists an integer y such that $x^2=8y+1$.

Problem 7.7 Find a counterexample for each statements below:

(1) For all real numbers a and b, if $b^2> a^2$, then $b>a$.

(2) The number n^3-n^2+n+4 is divisible by 5 for all positive integers n.

Problem 7.8 Prove that for all nonnegative integers a, b the difference of squares a^2-b^2 does not give a remainder of 2 when divided by 4.

Problem 7.9 Let a, b, c be three real numbers. Prove that at least one of $a(a+b+c)$, $b(a+b+c)$, $c(a+b+c)$ is nonnegative.

Problem 7.10 Establish the truth value of the following statements:

(1) $(\exists x\in \mathbf{Z})(\exists y\in \mathbf{Z})x^2\leqslant y^2$;

(2) $(\exists x\in \mathbf{Z})(\forall y\in \mathbf{Z})x^2\leqslant y^2$;

(3) $(\forall x\in \mathbf{Z})(\exists y\in \mathbf{Z})x^2\leqslant y^2$;

(4) $(\forall x\in \mathbf{Z})(\forall y\in \mathbf{Z})x^2\leqslant y^2$.

Problem 7.11 Let us define $f: \mathbf{N}\to\mathbf{N}$, $f(n)=n^2+n+41$ and observe that $f(0)=41$, which is prime, $f(1)=43$, which is also prime and $f(2)=47$, which is again prime. Is it true that $f(n)$ is prime for any nonnegative integer n? Find a counterexample!

Problem 7.12 For which positive integers n is it true that $n!>2^n$?

Problem 7.13 Prove that 6 divides n^3+11n for all nonnegative integers n.

Problem 7.14 Prove that 7 divides $2^{4n+2}+3^{2n+1}$ for all nonnegative integers n.

Problem 7.15 Prove that for every positive integer n,

$$1^2+2^2+\cdots+n^2=\frac{n(n+1)(2n+1)}{6}.$$

Problem 7.16 Prove that for every integer $n\geqslant 2$,

$$\frac{1}{n+1}+\frac{1}{n+2}+\cdots+\frac{1}{2n}\geqslant\frac{7}{12}.$$

Problem 7.17 Let a and r be real numbers, $r\neq 1$. Prove by mathematical induction that, for all nonnegative integers n,

$$a+a\cdot r+a\cdot r^2+\cdots+a\cdot r^n=\frac{a(1-r^{n+1})}{1-r}.$$

Problem 7.18 Prove that if x is a nonzero real number and $x+\frac{1}{x}$ is rational, then $x^n+\frac{1}{x^2}$ is rational for any positive integer n.

Problem 7.19 If we draw n straight lines in the plane, no three going through the same point, and no two parallel, how many regions do they divide the plane into? Prove your answer.

Problem 7.20 The Lucas sequence is a sequence of integers f_1, f_2, $\cdots$, f_n, $\cdots$ such that $f_1=1$, $f_2=3$ and

$$f_{n+1}=f_n+f_{n-1}$$

for all $n\geqslant 2$. Find the pattern for the remainder when f_n is divided by 3. Prove your answer. Is $f_{1\,000}$ divisible by 3?

Problem 7.21 The Fibonacci sequence is a sequence of integers g_1, g_2, $\cdots$, g_n, $\cdots$ such that $g_1=g_2=1$ and

$$g_{n+1}=g_n+g_{n-1}$$

for all $n\geqslant 2$. Prove that for all n,

$$g_n=\frac{1}{\sqrt{5}}(\alpha^n-\beta^n),$$

in which $\alpha=\frac{1+\sqrt{5}}{2}$ and $\beta=\frac{1-\sqrt{5}}{2}$.